Hormonal Chaos

The Scientific and Social Origins of the Environmental Endocrine Hypothesis

Sheldon Krimsky

The Johns Hopkins University Press
Baltimore and London

Printed in the United States of America on acid-free paper
9 8 7 6 5 4 3 2 1

The Johns Hopkins University Press
2715 North Charles Street
Baltimore, Maryland 21218-4363
www.press.jhu.edu

Library of Congress Cataloging-in-Publication Data will be found at the end of this book.
A catalog record for this book is available from the British Library.

ISBN 0-8018-6279-5

CONTENTS

FOREWORD

In this book Sheldon Krimsky traces the scientific roots of the environmental endocrine hypothesis and shows how independent lines of research arrived at similar conclusions. He reveals the role played by a small group of scientists in bringing about greater public awareness of the hazards of endocrine-disrupting chemicals.

We get an insider's look at how scientists debate the evidence, how the media report the issues of sperm decline and wildlife abnormalities, and how Congress addresses public concerns. Krimsky places the debate over endocrine-disrupting chemicals in the larger context of chemical regulation by critically examining the assumptions underlying our current policies and questioning the progress we are making in protecting our environment from the release of toxic substances.

The endocrine disrupter issue was at the center of my radar screen from the first day of my term (1993–99) as assistant administrator of the Environmental Protection Agency's Office of Prevention, Pesticides and Toxic Substances. Such was the sense of urgency attached to the issue that, one day in October 1993, EPA Administrator Carol Browner was asked to swear me into office on an urgent basis so I could testify before a hearing of a subcommittee of the House Government Oversight Committee about estrogens in the environment. A number of scientists testified at the hearing, and it was impressive to observe the level of interest of the members on both sides of the aisle. As a pediatrician, I was especially concerned about what I was hearing from scientists regarding the impacts of endocrine disrupters on developing fetal animals, both in the wild and in the laboratory. This testimony certainly raised my level of concern about the potential for effects in children and convinced me that the issue should become a priority for the EPA's chemical and pesticide programs. At the same time, I recognized that there were a number of gaps in our knowledge that would have to be filled before we could understand the regulatory implications of these new insights.

In August 1996, Congress enacted the Food Quality Protection Act and the Safe Drinking Water Act. Both bills included requirements for screening and testing chemicals for endocrine-disrupting properties. These statutes required screening for estrogenic chemicals in food and source drinking water and authorized the EPA to screen for other hormonal effects as well. This was landmark legislation. The Toxic Substances Control Act, enacted in 1976, gave the EPA broad authority to require testing of chemicals for safety, but very little testing had been carried out in the ensuing 20 years. This was frustrating for me in trying to set priorities. Years of administration and legal precedents, as well as the law itself, made it difficult to increase the pace at which essential hazard information was generated, even for commonly used chemicals. Moreover, decisions by past administrations had stripped both my office and EPA's legal staff of the resources that would support issuing testing requirements. Thus I viewed endocrine disrupter screening as the first real opportunity in 20 years to conduct serious, large-scale data gathering on chemicals in the United States (which numbered 85,000 in 1996).

Early in 1996, the Chemical Manufacturers Association and Theo Colborn of the World Wildlife Fund jointly approached me with an unusual request. Could the EPA explore the possibility of developing an endocrine disrupter screening program by bringing together the talent and resources of stakeholders in a consensus-based process? Over the next few months, it became clear to me that such a process could be successful. Nevertheless, there were many potential obstacles. I have already mentioned the resource limitations within the EPA. Just as challenging were the complexity of the scientific issues and the strict deadlines established by Congress. In addition, the issue of endocrine disruption had threatening—and emotional—overtones for industry and environmental groups alike. It was with very dubious (but,

I am thankful to say, bright, hardworking, and loyal) troops that I initiated the process of negotiating the Endocrine Disrupter Screening and Testing Program. The process was a success for a number of reasons, but primarily because so much groundwork had been laid beforehand and because of the caliber and dedication of the participants. The wonderful facilitation by the staff of the Keystone Center and the Meridian Institute and the hard work of employees of the EPA and other agencies were also critical to the program's success.

Hormonal Chaos weaves the science behind and popular responses to the environmental endocrine hypothesis into a fully integrated account that brings the disparate elements and opposing voices of this important environmental debate into sharp focus. I strongly recommend it as reading not only for those who are interested in the endocrine disrupter issue per se but also for students of environmental policymaking who are interested in the interplay between science and policy.

Lynn R. Goldman, M.D., M.P.H.

PREFACE

I first became aware of a controversy over chemicals that behave like human estrogens from conversations I had with two of my colleagues at the Tufts University School of Medicine in 1993. Carlos Sonnenschein and Ana Soto participated with me in a universitywide initiative to build an interdisciplinary center on the environment. I was curious about why two members of the Department of Anatomy and Cell Biology, whose work focused on cell proliferation, were interested in such a center. I subsequently learned about their accidental discovery in late 1989 that a chemical used in the manufacture of plastic laboratory devices had leached into their test cells, causing them to reproduce as if they had been exposed to a human estrogen. Sonnenschein and Soto told me of other scientists who had discovered links between hormonally active industrial chemicals and human and wildlife abnormalities. After a brief investigation into the subject, it became apparent to me that an important scientific hypothesis with serious environmental implications was in the making. This book discusses the events that have contributed new insights into and sparked new controversies over the effects that some chemical products of our industrial age are having on living organisms.

The book is divided into five chapters. Chapter 1 examines the scientific roots of the theory that chemicals in the environment can interfere with hormonal messages in living organisms (including humans), affecting reproduction and development, and also giving rise to disease. My historical inquiry includes a brief look at *Silent Spring*, Rachel Carson's classic study of the effect of synthetic organic pesticides on wildlife and humans. Carson was a member of the nature literati—a synthesizer and science communicator. She built a bridge across disciplines, made scientific findings accessible to a wide public audience, and created a model of lucid and literate scientific analysis that has become a classic in American environmental history.

Although Carson made some oblique references to the hormonal effects of chemicals, it took a very different set of events to clarify the mechanisms through which hormonal interference can result in disease and developmental abnormalities. This chapter explores the confluence of three principal scientific paths that resulted in the environmental endocrine hypothesis and describes the period of multidisciplinary cooperation among scientists during which a generalized explanation of the role of xenobiotic endocrine disrupters was formulated.

Chapter 2 gives an account of how this scientific hypothesis gained public notoriety. A powerful film, a provocative book, congressional hearings, pathbreaking legislation, and plentiful media attention all amplified and popularized the risks raised by concerned scientists. Yet there were counterforces as well. Scientific skeptics became vocal. Industrial groups were poised to protect product and sector interests. Journal editors were cautious about giving too much weight to a scientific conjecture. The voices of concern and skepticism, carried into the public arena through the Internet and print and electronic media, created societal constituencies for investigating the effects of chemicals that had until then been poorly studied and understood.

Chapter 3 examines the response to the hypothesis by the scientific community. It explores the areas of scientific uncertainty within a generalized causal framework and discusses the responsibility of scientists in the face of provocative but nondefinitive evidence of human health effects. The environmental endocrine hypothesis is set in the context of other ethical dilemmas raised by the dissemination of and societal response to early research findings that posit a public health risk.

Chapter 4 addresses the challenges facing the policy sector. How does government respond to demands for regulating chemicals under new criteria when there are so many gaps in our knowledge of their effects? Endocrine-disrupting synthetic chemicals may behave contrary to principles of traditional toxicology, adding to the complexity of applying the current laws and regulations intended to safeguard human and environmental health. This chapter also explores the complications to which additive and synergistic effects of chemicals give rise, as well as the prospect that extremely low levels of exposure to endocrine disrupters can have significant consequences for the developing fetus that are not manifest for years or even decades after birth. The social response to the environmental endocrine hypothesis is explored through the alternative ethical norms of the newly framed "precautionary principle" and the more conventional standard of demonstrated risk. This chapter takes us through the U.S. policymaking process charged with formulating a testing strategy for determining which chemicals can be classified as endocrine disrupters. In Chapter 5, the conclusion, I discuss the degree of consilience reached by the scientific community on the endocrine hypothesis, the remaining areas of contested knowledge, and the role of the hypothesis in expanding the assessment of chemical hazards.

This work was supported in part by a grant (SBR-94/2973) from the Ethics and Value Studies Program of the National Science Foundation, with additional support provided by Tufts University. I wish to acknowledge the contributions of Roger Wrubel to the background research on behavioral and neurophysiological effects of endocrine-disrupting chemicals, to the analysis of the scientific review articles discussed in Chapter 1, to the discussion of the media in Chapter 2, and to the compilation of reviews of *Our Stolen Future* in Chapter 3.

Kelly Morgan and Jennifer Patrick surveyed the Internet for endocrine disrupter sites, which they classified and analyzed. Ms. Patrick provided invaluable help in locating studies, in checking citations, and in preparing tables and figures. My special appreciation goes to Dianne Dumanoski, who twice taught a seminar with me on the subject matter of this work and who has provided thoughtful comments and advice throughout its development. I wish to thank my colleagues Carlos Sonnenschein and Ana Soto, who gave me continuous support during this project. I am greatly appreciative of the following individuals who read and commented on sections of the manuscript: Dianne Dumanoski, Anthony Maciorowski, John Peterson Myers, Ingar Palmlund, Rosalind Rolland, and Ted Schettler. I am also indebted to the more than two dozen people who agreed to be interviewed for this book. All of your names may not appear in its pages, but each of your insights and perspectives was important in its development.

Finally, I wish to thank Ginger Berman, my editor at the Johns Hopkins University Press, for her support and suggestions throughout this project, and Princeton Editorial Associates, for their superb editorial work on the manuscript.

Hormonal Chaos

CHAPTER 1

Scientific Developments

Early in my academic career members of my department at Tufts University supervised several studies of communities embroiled in struggles over toxic pollution. One of my colleagues chose a site in Woburn, Massachusetts, a northern Boston suburb where a cluster of childhood leukemia cases had been discovered in a section of the city bordering an industrial site. I chose to supervise an investigation of groundwater pollution in Acton, Massachusetts, another Boston suburb, where residents felt assaulted by foul air from industrial smokestacks and feared that toxic chemicals, summarily dumped into open lagoons, had leached into the town's drinking water.

In both communities, residents organized and educated themselves, first to gain information and then to seek justice. In both cases, federal suits were filed against polluters, and several town wells were closed. Ironically, the two communities even had one multinational chemical company in common. Yet beyond these similarities, the cases began to diverge. Woburn was listed as a national Superfund site. It inspired television documentaries and books, the most famous of which is Jonathan Haar's *A Civil Action,* which was the basis of a motion picture starring John Travolta and Robert Duvall, released in December 1998.

Meanwhile the pollution of Acton's groundwater remained a regional story, despite a major lawsuit filed by the Department of Justice, charging violations of the Resource Conservation Recovery Act. Notwithstanding the differences in the nature of the pollutants affecting these communities, the most poignant distinction between these cases was the cancer clusters discovered in one community and missing in the other.

For nearly half a century, the fear of toxic chemicals has been largely synonymous with the fear of cancer. America's obsession with cancer—the dread disease—has been the dominant lens through which the study of toxic chemicals has been viewed. States have established cancer registries. The federal government has enacted a significant body of legislation responding directly to cancer risks in the environment, in food, and in the workplace.

The scientific community responded to the societal war against cancer with a theory of cancer causation, the somatic mutation theory of carcinogenesis, which quickly assumed dominance in public discussion. Simply stated, the theory holds that certain chemicals, viruses, or radiation acting on cells can cause changes (mutations) in the cells' DNA, and that these changes cause the cells to become cancerous.

Cancer is certainly not the only adverse health effect of industrial chemicals, but it has largely eclipsed other diseases and reproductive effects as an object of public concern and scientific research. Similarly, the somatic mutation theory of carcinogenesis has left little room for other theories of chemically induced disease to compete for research funding. However, in the last few years a new theory of environmental disease has emerged that explores a variety of human and animal abnormalities that are not explained by or investigated within this dominant cancer paradigm. The guiding concept of this new paradigm is that some chemicals can interfere with the body's natural hormones.

For almost two decades scientists from a variety of disciplines had been following seemingly independent research paths that were eventually to be connected through a bold and unorthodox hypothesis. The hypothesis, which I refer to as the *environmental endocrine hypothesis,* asserts that a diverse group of industrial and agricultural chemicals in contact with humans and wildlife have the capacity to mimic or obstruct hormone function—not simply disrupting the endocrine system like foreign matter in a watchworks, but fooling it into accepting new instructions that distort the normal development of the organism. Mostly synthetic organic chemicals, these compounds have been implicated in more than two dozen human and animal disorders, including reproductive and developmental abnormalities, immune dysfunction, cognitive and behavioral pathologies, and cancer. Some of the postulated effects of endocrine-disrupting chemicals (or, for short, endocrine disrupters) have been corroborated in wildlife and laboratory studies.

If the hypothesis is confirmed for human effects, science will have discovered an important new etiology of environmental disease. From the standpoint of human pathology, the environmental endocrine hypothesis could turn out to be the most significant environmental health hypothesis since the discovery of chemical mutagenesis. It has the potential to change radically the way we think about the contribution of environmental factors to disease by shifting the focus from cancer and acute toxicity to the reproductive, neurophysiological, and developmental effects of chemicals. From the standpoint of global environmental effects, the hypothesis may be the most important finding since the discovery that chlorofluorocarbons deplete the protective ozone layer in the atmosphere.

The implications of the hypothesis for law and regulation are also profound. Currently, dozens of chemicals, including pesticides and commonly used plasticizers, exhibit hormone-like or hormone-antagonistic behavior in laboratory assays and animal experiments. Reproductive declines in wildlife have been linked to hormone-disrupting chemicals that have been introduced into the animals' habitats. The estrogen effects of industrial chemicals and pesticides were the first to be discovered. Reptiles, birds, and fish in their natural habitats have been estrogenized by chemical exposures. Some of the chemicals that exhibit estrogenic effects are produced in the United States in quantities far in excess of several hundred million pounds per year. Other chemicals, although not explicitly estrogenic, exhibit antiandrogenic (obstructing androgens) or thyroid-disrupting behavior that is also capable of affecting the endocrine system's hormonal balance in living organisms. Once exposed, humans and wildlife may accumulate endocrine disrupters in serum or tissue. Even small traces of these substances may prove harmful, because their effects may be additive when combined with those of traces of other unmetabolized synthetic hormone disrupters that have taken up residence in the organism.

Although certain components of the environmental endocrine hypothesis are still being debated among scientists, industrial nations have already begun considering how to regulate a class of chemicals that is currently ill-defined yet ubiquitous. Governments will be faced with the challenge of assessing the health and environmental effects of numerous individual chemicals as well as the cumulative effects of multiple exposures to many different kinds of chemicals. Once this information becomes available, nations will have to decide on acceptable levels of these endocrine-disrupting substances, which have become so pervasive in the products and processes of modern life.

The environmental endocrine hypothesis evolved slowly but steadily from evidence compiled by scientists from a multiplicity of disciplines who did not have its broad outlines in mind as they were engaged in their own studies. Their work became identified with the hypothesis only in the late 1980s when Theo Colborn, a scientist working for the nonprofit Conservation Foundation, proposed a unified explanation for reproductive and developmental abnormalities in wildlife populations inhabiting the Great Lakes. Colborn eventually extended the hypothesis to all living organisms, discovering in it a powerful organizing principle for understanding hormone-mediated abnormalities within the human population.

The scientific paths leading to the environmental endocrine hypothesis may be traced to three independent research programs. First was the discovery of the intergenerational health effects of the synthetic estrogen diethylstilbestrol (DES), a drug administered to pregnant and post-menopausal women for 20 years. Since it is not produced naturally within the body, DES is called a xenobiotic (a biological agent foreign to the organism) estrogen or xenoestrogen. Second was a sizable body of field and laboratory studies linking wildlife reproductive disorders to chemical effluents from industrial and municipal waste and agricultural pesticides. And third was research investigating a postulated global decline in the quality and quantity of human sperm. The common thread connecting these independent research paths was recognized by Colborn, who, while in search of an integrated theory, built bridges across many scientific disciplines.

The social factors that play a role in the development of a new scientific hypothesis and in promoting public awareness of it are very much a part of this story. It was, in the final analysis, an outsider, a woman of unusual vision and determination, who convinced the scientific community to regard the environmental endocrine hypothesis as a credible explanatory framework for a diverse assortment of environmental phenomena. In the historical account of this remarkable episode in the health sciences there will be ample opportunity to examine the social context of scientific discovery. In particular, how does an outsider to the established research communities, working for a nongovernmental organization, bring about a paradigm shift in the etiology of environmental disease—one that has captured the attention of the world's scientific and policy communities? Theo Colborn has been compared to America's preeminent environmental heroine, Rachel Carson, who also led a revolution in the social and scientific perception of the environmental effects of industrial chemicals.

Following the trail of scientific discovery is a little like tracking a family genealogy. No matter how far back you go, there are always deeper roots. The choices of where to begin the family tree and how many branches to explore are practical ones, limited by the commitment of time, the allocation of resources, and the existence or availability of records. Similarly, the scientific story of the environmental endocrine hypothesis has some practical points of departure, but, as with the recounting of all scientific discoveries, any choice of origin always tends to devalue the preceding work.

I choose to begin the story by introducing a British scientist, Sir Edward Charles Dodds (1889–1973), a professor of biochemistry at the Middlesex Hospital Medical School at the University of London and fellow of the Royal Society. In 1959, at its 282nd convocation, the University of Chicago awarded Dodds one of several honorary degrees he received during his career. The citation called him "an ingenious biochemist noted for his contributions to endocrinology which have relieved untold amounts of human suffering." Dodds had won international acclaim for his synthesis of synthetic estrogen, which he called stilbestrol, in 1938. In preparation for that achievement, he had begun investigating the estrogenic properties of organic compounds during the early 1930s. Dodds discovered that human and animal hormonal function could be induced by substances found in nature and by synthetic variants of those substances. He believed that a specific portion of an estrogen molecule was responsible for estrogenic activity, and he began isolating likely suspects. To test these substances, Dodds and his colleagues used ovariectomized rats: rats that had had their ovaries removed so that they could not produce the hormone naturally. An estrogenic effect would be confirmed when the rats' vaginal cells responded to the external chemical as if they had been exposed to natural estrogen. The true point of departure for the story of the environmental endocrine hypothesis is found in the discovery by Dodds of the synthetic estrogen diethylstilbestrol.

Building on Dodds's work, two scientists in the Department of Zoology at Syracuse University published a paper in 1950 that received little recognition, but which, in retrospect, serves as the second branching point for the trail of the endocrine hypothesis.* Howard Burlington and Verlus Frank Lindeman chose 70 male chicks, divided into 30 controls and 40 experimental animals, to investigate the developmental effects of a pesticide that had been synthesized in 1938. The experimental animals were given daily injections of a purified form of the insecticide dichlorodiphenyl trichloroethane (DDT) in a solvent of chicken fat, while the controls received injections of the chicken fat alone. The results were striking. The experimental animals had smaller testes and showed arrested development of secondary sex characteristics compared with the controls. The authors wrote: "The effects noted here are such that they might easily be duplicated by the administration of an estrogen. It seems, therefore, that the possibility of an estrogenic action of DDT is at least worthy of consideration. In speculating along these lines, it is interesting to note the degree of similarity between the molecular configuration of DDT and certain synthetic estrogens, especially diethylstilbestrol."[1]

The critical connection made between DDT and DES in 1950 was to remain dormant for over 20 years. Yet these two synthetic chemicals—one a popular pesticide and the other a widely used human and animal drug—are a key to understanding how several scientific paths eventually converged in the discovery of the environmental endocrine hypothesis. The next branching point in this scientific genealogy takes us to Rachel Carson, who drew attention to the hazards of pesticides generally, but especially to DDT.

Roots of the Endocrine Hypothesis in *Silent Spring*

Rachel Carson has been anointed by the media and environmental historians as the patron saint of the modern environmental movement. There is justice in this accolade. Her book *Silent Spring* has been read and reread for its prophetic insights, for its telltale clues to new environmentally mediated diseases, and for its enduring message that what we humans do to nature eventually affects us. Carson, who achieved notoriety as a nature writer, set out in *Silent Spring* to demonstrate the connections between changes taking place in wildlife ecology and new cases of human disease. She conjectured that rising cancer rates could be attributed to "countless new cancer-causing" chemicals introduced into

*The reference to the Burlington and Lindeman paper was made opaquely in *Silent Spring,* and it was rediscovered by Dianne Dumanoski while she was researching *Our Stolen Future.*

the environment through modern methods of manufacturing and agriculture. The toxic effects of these chemicals, she wrote, were beginning to manifest themselves in the general population, in people completely unaware of their own exposure. Carson believed that "no longer are exposures to dangerous chemicals occupational alone; they have entered the environment of everyone—even children as yet unborn."[2]

Carson devoted much of her fourteenth chapter, which she titled "One in Every Four," to the concern over chemically induced cancers. Indeed, while she was writing the book, Carson was battling her own affliction with cancer. She considered the evidence she had amassed on the human health effects of pesticides as circumstantial, but nevertheless impressive. Carson is all too often associated quite narrowly with DDT, but she had a much broader perspective on the dangers chemicals posed to human health. She cited insecticides, weed killers, fungicides, plasticizers, medicines, clothing, and insulating materials as potential cancer-causing substances.[3]

Mutagenicity—the alteration of a cell's DNA by chemicals or radiation—was identified by Carson as the most likely cause of cancer. It was the dominant paradigm of cancer etiology at the time. But she also considered other routes to the class of diseases we call cancer, ones that were connected with reproductive abnormalities and disorders of the endocrine system—that complex of glandular organs that produce a myriad of messenger hormones that regulate so many of our vital bodily functions. Her writings in *Silent Spring* are precursors to the current hypothesis of xenoestrogens and, more broadly, endocrine disrupters. Carson also believed that some cancers arose out of a multistage process, thus making it extremely difficult to isolate a single cause: "The road to cancer may also be an indirect one. A substance that is not a carcinogen in the ordinary sense may disturb the normal functioning of some part of the body in such a way that malignancy results. Important examples are the cancers, especially of the reproductive system, that appear to be linked with disturbances of the balance of sex hormones; these disturbances, in turn, may in some cases be the result of something that affects the ability of the liver to preserve a proper level of these hormones."[4]

Carson's insights into the risks these chemicals posed for humans were based on observations in wildlife. Birds were not reproducing; therefore the endocrine system was implicated. She posited a connection among chemicals, cancer, hormones, and reproduction: "The sex hormones are, of course, normally present in the body and perform a necessary growth-stimulating function in relation to the various organs of reproduction. But the body has a built-in protection against excessive accumulations, for the liver acts to keep a proper balance between male and female hormones . . . and to prevent an excess accumulation of either."[5]

Carson believed that organ damage resulting from exposure to foreign chemicals could result in elevated levels of male or female hormones. Estrogens that build up to abnormally high levels, she surmised, could result in carcinogenic effects or otherwise impair the liver. She cited evidence that certain cancers were associated with high estrogen levels, a correlation that is today well established. Thus, she suggested, chemicals that cause damage to the liver may be sufficient to interfere with the process of estrogen elimination, resulting in excess levels of endogenous estrogen and in an overestrogenized body that would be at increased risk of cancer. She went on to discuss the problem of exogenous estrogens: "Added to these are the wide variety of synthetic estrogens to which we are increasingly exposed—those in cosmetics, drugs, foods, and occupational exposures. The combined effect is a matter that warrants the most serious concern."[6]

Carson stopped one sentence short of positing an estrogen-mimicking function of synthetic chemicals whose estrogenic or hormone-disrupting properties were hidden from us. It was not until the mechanism of action of hormones became better understood that it was possible for scientists to accept the role of foreign substances as hormone mimics or antagonists. Although it became widely acknowledged years later that Carson's chemical nemesis, DDT, exhibited estrogenic effects—a result first recorded in the scientific literature by Burlington and Lindeman (1950)—it was a synthetic human estrogen that first brought serious scientific attention to the role of xenobiotic chemicals in the human and animal endocrine system.[7]

DES as a Xenoestrogen

Research on DES, the first synthetic estrogen—including investigations into the etiology of the iatrogenic (treatment-caused) diseases it has conferred on women and men—was one of the principal tributaries of the environmental endocrine hypothesis. DES, while not chemically similar to natural estradiol, in some respects mimics the action of that hormone, which causes the lining of the uterus to thicken during the beginning of the menstrual cycle.

The case of DES played two important roles in providing foundational knowledge to support the environmental endocrine hypothesis. First, DES is a synthetic estrogen, and therefore the adverse effects of DES treatments on human health focused attention on it and eventually on other types of exogenous estrogenic compounds, both drugs and environmental chemicals. Second, animal models that had been developed to understand the mechanism of DES action could be generalized to the study of other xenoestrogens. In 1987 McLachlan and Newbold wrote that the exposure of animals to DES in a laboratory setting provided a "possibly exaggerated model for what might be seen with much weaker estrogenic xenobiotics, many of which are common features in our environment."[8] The study of the role of endogenous estrogens in reproduction and sexual development led naturally to investigations of the synthetic estrogens that were used to augment low levels of estrogen production or study the benefits of increased estrogen production. Research on synthetic estrogens as therapeutic drugs led researchers to consider that synthetic industrial and natural plant chemicals might also exhibit estrogenic effects. As evidence of the xenobiotic estrogenicity of industrial and agricultural chemicals grew, the scientific and historical importance of the DES episode became more widely recognized.

After Dodds's discovery in 1938, the first and most widely prescribed synthetic estrogen, DES, was manufactured from coal tar derivatives. Synthetic estrogenic substances were subsequently marketed in Europe and North America under more than 400 different trade names.[9] When first synthesized, DES was known to have two important properties: it was highly estrogenic (even more potent than estradiol) and highly carcinogenic. A year after DES had first been synthesized, it was made available in the United Kingdom, France, Germany, Sweden, and the United States for a variety of treatments associated with women's health and subsequently for the treatment of prostate cancer.[10]

By the early 1940s, DES use in the United States had expanded to include applications in commercial agriculture. The first tests of the chemical as a growth stimulant in beef cattle were conducted in 1947 at the Purdue University Agricultural Station. Published reports of DES in beef cattle feed supplements first appeared in 1954 and 1955 from experiments conducted at Iowa State University.[11]

In 1941 the Food and Drug Administration (FDA) approved a new-drug application for DES as a treatment for menopausal women. Subsequently, the FDA extended the use of DES to a variety of conditions associated with pregnancy. Physicians prescribed it to prevent spontaneous abortion and stillbirths, to suppress lactation in women who wished to bottle-feed, and as a postcoital contraceptive.[12] Between the late 1940s and 1971, physicians prescribed DES to as many as 3 million pregnant women in the United States.* However, studies subsequently found that DES was ineffective as a treatment for the disorders or for the preventative functions for which it had been prescribed.[13]

After DES had been shown to be carcinogenic in test animals and residues of it were found in poultry, the FDA suspended its use in poultry in 1958—but permitted continued use of the drug in cattle and sheep. The drug amendments to the Food, Drug and Cosmetic Act, passed in 1962, required the FDA to demand proof of efficacy from drug companies for all drugs introduced between 1938 and 1962. And so DES was reevaluated based on the new standards.

During his tenure at the Harvard Medical School and Massachusetts General Hospital, Arthur Herbst discovered a relationship between vaginal cancer and DES. In 1971 he and his colleagues reported eight cases of clear cell adenocarcinoma of the vagina in women under 20, an extremely rare condition for this age group. The mothers of all but one of the women had been treated with DES during pregnancy.[14] The effects of in utero exposure to DES were also observed for male progeny: men exposed to DES during gestation were more likely to exhibit abnormalities in sex organs, reduced sperm count, and lower quality of sperm compared with controls.[15]

These results brought to a head the protracted process of withdrawing DES from commercial use. In 1971 the FDA required the labeling of DES to contraindicate its use in pregnancy, and in 1972 it withdrew all approvals of DES for animal use. It took five more years of litigation before the courts eventually upheld the FDA's ban on the use of DES in animals. At the peak of its use as a growth promoter in the 1960s, DES had been administered to nearly 30 million cattle in the United States.

Although cancer in the genital tract was a central theme of much of the animal research in the 1970s, there were already glimpses in the DES literature of effects beyond cancer. Examples of other effects are found in Herbst and Bern's 1981 edited volume: "Perinatal exposure of male laboratory animals to DES yields results that are relevant to similarly exposed humans. Cryptorchidism [undeveloped testes], epididymal cysts, hypoplastic testes, sperm abnormalities, and prostatic inflammation have been described both in laboratory animals and humans. . . . Marked alteration of fertility in the male mouse has been noted following antenatal exposure to hormones and alterations in semen have been reported in humans."[16]

*Palmlund documents the wide-ranging uses of DES on women before and during pregnancy, during childbirth, and in postpartum and neonatal care.

During the 1970s, John McLachlan emerged as a pivotal figure in applying the results of DES research to investigate other DES-like effects. McLachlan had begun thinking about the role of estrogens even prior to his DES work—as far back as 1968, while searching for a dissertation topic. His thesis adviser was a gynecologist and research scientist who was interested in the transfer of drugs into uterine secretions and their subsequent effects on the developing fetus. McLachlan completed his doctorate in pharmacology at the George Washington University in 1971. For his dissertation he studied the transfer of nicotine from the maternal circulation to uterine secretions and ultimately into the blastocyst (the multicellular embryo whose cells have not yet differentiated). His research involved administering an "environmental cocktail" consisting of nicotine, caffeine, salicylate, and DDT to pregnant animals. The compound with the most lasting effects on the uterus and the blastocyst, according to McLachlan, was DDT.

At about that time, McLachlan began reading reports from wildlife studies describing the weak estrogenicity of DDT. He also learned from Herbst's 1971 paper that DES was a transplacental carcinogen (one that crosses the placenta from the mother to the developing fetus).[17] He considered this the opportune time to develop an animal model to study the first known xenoestrogen linked to cancer and then to apply this model to the investigation of other synthetic substances.

McLachlan did his postdoctoral work at the National Institutes of Health. It was there in 1971 that he began looking at the long-term effects of DES as an "estrogenic surrogate" for DDT.[18] He chose DES as a model to study DDT because it was inexpensive and synthetic, and bore some structural resemblance to the pesticide. In 1973 he accepted a position at the Laboratory of Reproductive and Developmental Toxicology at the National Institute of Environmental Health Sciences (NIEHS) in Research Triangle Park, North Carolina. At NIEHS he continued his work on DES, seeking to learn whether it was a carcinogen or teratogen (an agent capable of causing developmental disorders) and to elucidate its mechanism of action. By 1976 he had been made head of the Developmental Endocrinology and Pharmacology Section of the laboratory.

McLachlan organized the first symposium on estrogens and the environment in 1979 and edited the published volume of symposium papers (see Appendix A).[19] He pursued the question of whether DES was a good chemical model for investigating DDT. By the mid-1970s, both compounds had made it to the top of America's list of most suspect chemicals. This meant that new federal funds were available for studies of their health effects. While continuing his investigations into DES, McLachlan contacted physicians to determine whether their clinical observations were consistent with the findings from his experiments in mice. His interest in environmental estrogens began to develop from his study of DES's mechanism of action and its pervasiveness in the environment.

At about the time of the 1979 "Estrogens in the Environment" meeting, McLachlan learned that considerable quantities of DES had been introduced into the food supply. Prior to 1977, the cattle industry had used over 13 tons of DES a year as an additive in beef. McLachlan invited to the meeting scientists who could discuss the persistence of DES in the environment and who could apply chemical structure modeling to investigating the properties that lead to estrogenicity. This process involves looking for structural similarities among chemicals to predict or account for their estrogenic effects on biological systems. It soon became evident to McLachlan that biochemists would not be able to solve the estrogen problem through a structural analysis of the chemicals. The various known estrogenic compounds were not homologous (structurally similar). There was no telltale hydroxyl or methyl ring, no unique configuration of carbon atoms. The identification of other estrogenic substances would have to be made on the basis of the role that they played in the biological system.

Seeking to understand the mechanism through which estrogens could induce cancer, in 1980 McLachlan and his collaborators continued to use the DES mouse model to study whether synthetic hormones cause events similar to those produced by endogenous hormones.[20] They considered the mouse particularly well suited to study the mechanisms of induction and expression of epithelial cancers by estrogens. As one of the most potent of all the estrogens, the synthetic hormone DES provided an extreme example of the biological effects of a xenoestrogen.

McLachlan also used the DES mouse model to explore the role of estrogens in sex differentiation. In their collaborative paper, McLachlan and Retha Newbold stated that, based on their DES-mouse studies, "exposure to exogenous estrogens during fetal development will profoundly alter sexual differentiation."[21] They cited a number of chemicals of diverse structure that shared one functional characteristic—estrogenicity. They provided examples drawn from phytoestrogens, mycotoxins, polycyclic aromatic hydrocarbons, and chlorinated hydrocarbons. The DES studies clarified the mechanism of action as well as the reproductive effects on animals. A number of scientists began investigating the metabolism of DES as well as the estrogenic activity of industrial chemicals. Generalizations from DES to other estrogenic compounds began to appear in the literature. As new assays for estrogenicity were developed and more substances were tested, the concern over xenoestrogens widened into a broader set of hypotheses.

Looking back, McLachlan described the period up to 1985 in the pursuit of environmental estrogens as a lonely one. The response he received to articles sent to mainstream journals was invariably that the pursuit of xenoestrogens was a kind of "funky science."[22] Reviewers considered the work metaphysical, pointing out that these compounds weren't really hormones. According to McLachlan, his detractors claimed he was stretching the limits of endocrinology and that his work was more like toxicology. He himself characterized his research as crossing the boundaries of endocrinology, developmental biology, and toxicology without fitting neatly into any one of the disciplines. He and his colleagues were creating their own branch of science, and it would take some years before it became accepted. At the time, the Endocrine Society (the main professional association of endocrinologists) would hold no special meetings on the subject of xenoestrogens. On the other hand, toxicologists were not thinking of these compounds within the confines of their discipline because traditional dose-response mechanisms did not apply to them. For toxicologists, the fundamental principle of chemical action in a biological system is that potency is linked to dose. The higher the dose, the greater is the expected effect. However, in dealing with hormones, small quantities might yield an effect, whereas large quantities of the same compound might shut the system off entirely, producing no effect.

According to McLachlan, the publication of his seminal coauthored paper in *Molecular Pharmacology* on the estrogenicity of DES and its analogues had established the tools and probes for the further study of xenoestrogens and helped place the research on firm scientific ground.[23] The controversy over the use of DES in pregnancy, to treat postmenopausal symptoms, and as an animal growth promoter set the stage for the first comprehensive study of a xenoestrogen. The connections between DES and industrial compounds like DDT, kepone, and polychlorinated biphenyls (PCBs) were beginning to be made at the NIEHS. The first symposium on estrogens in the environment in 1979 included papers on the effects of environmental hormone contamination in Puerto Rico and estrogens in food products. A new set of tools, including biochemical probes used for the study of DES metabolism, soon became available for investigating other substances that were suspected of being or had been shown to be estrogenic. Advances were made in the study of the molecular basis of hormonal action, including discoveries about hormone-receptor binding.

The body's hormones fulfill their vital roles as biochemical messengers that regulate metabolism and growth through a special class of molecules called *hormone receptors.* Scientists were in hot pursuit of the mechanism through which certain chemicals were able to activate hormone receptors and thereby promote biological activity. Initially it was thought that each hormone was uniquely matched to a specific receptor that resides either inside or outside the cells of specific tissues. Once the hormone (ligand) attaches itself to its "unique" receptor, this complex molecule can enter the nucleus of the cell, where it activates genetic processes that lead to the production of proteins (transcription). However, this lock-and-key model of hormone-receptor activation (as it is popularly known) had to be modified when it was discovered that foreign substances could bind to the receptor and mimic the biological action of a natural hormone. Moreover, two or more molecules that bind to the same receptor may not share the same chemical structure. Following the lock-and-key metaphor, we would say that each lock (receptor) may have multiple keys (molecules that activate the receptor), and that, even though the shapes of the molecules appear quite distinct, two such molecules can unlock the same receptor and thus may exhibit the same or similar hormonal properties.

As scientists continued to learn more about the molecular mechanism of hormone action, they discovered many variations in the model. Some foreign substances might attach to the hormone receptor but not activate a genetic response. Instead, these substances take up spaces that would otherwise be filled by natural hormones. Thus, these foreign molecules can depress the vital role that the body's hormones are designed to play by preventing the natural hormones from attaching to their receptor sites. In other words, these rogue "hormones" can muck up the works.

During the 1980s the idea of environmental hormone disrupters was scarcely acknowledged within the scientific community. The journal *Environmental Health Perspectives* and its parent agency, the NIEHS, provided much of the early support for papers and symposia on xenoestrogens. Researchers at NIEHS, especially McLachlan, brought together scientists from diverse disciplines who were able to share what they had learned about DES and so facilitate the investigation of other xenoestrogens. In 1985 McLachlan organized a second conference on estrogens in the environment, which focused on the effects of chemicals on development.

Cadbury sums up the impact of McLachlan's early work: "Over the years he had written several books and had presented his ideas at international conferences. Although his work was appreciated within narrow scientific circles, where he was admired as a pioneer of a new approach to environmental hazard and as the 'father' of the field of environmental estrogens, the work did not hit the headlines. He published his ideas in the leading scientific journals and, like many scientists, never courted the popular press. The press, in turn, appeared oblivious of the wider significance of these studies."[24] Meanwhile, in another corner of science, wildlife biologists, pursuing questions about reproductive failure and developmental abnormalities, soon began asking questions similar to those posed by the investigators studying the human endocrine effects of DES, DDT, and other synthetic chemicals.

Wildlife Revelations

Anthropogenic explanations for declining wildlife populations have been extensively studied over the past 50 years. On a global scale, overexploitation of wildlife, habitat destruction, and the swelling of the human population are considered among the most important causes of species decline. Pollution in the form of agricultural chemicals, industrial effluents, landfills, and acid rain can destroy local wildlife populations either directly or indirectly by contaminating their food sources, resulting in the loss of suitable habitat necessary for maintaining biodiversity. However, pollution has been assumed in most cases to be but a minor cause of global wildlife decline and extinction. For example, the International Union for Conservation of Nature estimates that of the mammals and birds of Australia and the Americas classified as threatened, pollution represents the primary hazard in less than 5 percent of cases. In his highly acclaimed 1979 account of species extinction, *The Sinking Ark,* Norman Myers devotes a mere two pages to pollution as a source of species decline.[25]

The first significant claim that new classes of organic chemicals might have global impacts on wildlife came with the publication of *Silent Spring* in 1962. In the 1970s it was learned that certain pesticides had been found in the tissues of wildlife in the remotest parts of the world. It was also during that period that scientists discovered that the buildup of organic chemicals through the food chain was compromising the ability of predatory birds to produce offspring owing to eggshell thinning. More recently there has been scientific conjecture that outbreaks of disease in widely separated populations of marine mammals might be due to the impairment of the immune system of these animals resulting from their chronic exposure to organic pollutants. Scientists are confirming the observations of Rachel Carson years ago: the higher these marine organisms are on the food chain, the greater will be the concentration of the organic pollutants in their bodies—a process called *biomagnification*.

In this historical context of mounting evidence that environmental chemicals were affecting wildlife, the Conservation Foundation in Washington, D.C., and the Institute for Research on Public Policy in Ottawa, Canada, convened a team of experts in 1987 to examine the environmental condition of the Great Lakes basin. A small part of the overall study assessed the impact of chemicals on wildlife in the region. The 1990 book that grew out of the study, *Great Lakes, Great Legacy?*, asserted that chemical contamination was associated with the widespread reproductive failure in wildlife within the watershed, particularly that of fish and fish-eating birds and mammals. Although the concern of environmental regulators and wildlife toxicologists had previously been centered on acute toxic effects and cancer-causing consequences of xenobiotics, this new analysis identified other harmful end points.

One of the volume's contributors, Theo Colborn—moved by the wildlife as well as the human health data uncovered during her research for the book—continued to explore, with an ever-widening lens, additional studies of relevance to this theme. Although she was a newcomer to many fields of science, Colborn, a generalist and synthesizer, must be largely credited with the broad conceptualization and early dissemination of the idea of endocrine-disrupting chemicals in the environment. Starting from wildlife studies, she advanced the outlines of a new theory of environmental disease that encompassed the work of many other disciplines. We now turn our attention to the unique role and contributions of this outsider-scientist, who became, if not quite an insider, a respected and sought-after authority, a consensus builder, a facilitator of crossdisciplinary science, and standard-bearer of the environmental endocrine hypothesis.

Construction of a Science Constituency

Until the mid-1980s, the multidisciplinary scientific meetings that focused on environmental endocrine disrupters were centered around human health factors affecting growth and development. The causal agents included the synthetic estrogen DES and some industrial chemicals, particularly DDT and the family of over 200 PCBs. Interest in animals focused primarily on their roles as research models; as vectors that carried xenobiotics through the food supply, such as DES-fed livestock; and as sentinels, such as the DDT-exposed birds that manifested exposure through eggshell thinning. Ecological research on wildlife toxicology of the type that Rachel Carson had discussed in *Silent Spring* was continuing quite independently. Animal and in vitro studies demonstrated the estrogenic activity of insecticides such as kepone, mirex, dieldrin, and aldrin.[26] But it wasn't until Theo Colborn, a pharmacist turned zoologist, entered the scene in 1987 that the synthesis of wildlife ecology and human health studies began to take form.

Colborn was a pre-depression-era baby who had graduated from Rutgers University in 1947 with a major in pharmacy. She married Harry Colborn, also a pharmacist, and took up practice in his family drugstore in Newton, New Jersey. In 1962 the Colborns sold their business and purchased a sheep farm in western Colorado.

An avid birdwatcher, Colborn soon became an active environmentalist, participating in local campaigns to protect watersheds. She spent a summer doing field research at the Rocky Mountain Biological Laboratory and was awarded an M.A. in science in 1981 from Western State College of Colorado. Focusing on freshwater ecology, Colborn studied the transport of metals from mines to neighboring streams by measuring metal loadings in the skeletons of aquatic invertebrates. For her master's thesis Colborn studied aquatic insects as bioindicators of cadmium and molybdenum deposition in water.[27] After completing her degree, she was accepted into an interdisciplinary doctoral program at the University of Wisconsin at Madison in 1982. She continued her research, making insect biomarkers of heavy metals in aquatic systems the subject of her doctoral dissertation, and completed her Ph.D. degree in 1985 at the age of 58.[28]

In the summer of 1985 Colborn began a two-year congressional fellowship with the now-defunct U.S. Office of Technology Assessment, working on issues related to the reauthorization of the Clean Water Act. When her fellowship ended in 1987, she took a position with the Conservation Foundation, where she compiled much of the background research on toxic chemicals in the Great Lakes—a region of great concern among members of the foundation's advisory board.

Colborn worked with an interdisciplinary team that included policy specialists at the Conservation Foundation and the Institute for Research on Public Policy in Ottawa. She began an extensive literature search on the health of the ecosystem, including animals and humans, that took her through more than 2,000 scientific papers and 500 government documents. She arranged their data in a matrix highlighting the problems documented in the scientific literature for a variety of Great Lakes fauna (see Table 1). Of the 14 species identified as being in population decline, she found signs of reproductive abnormalities in 11 species and tumors in only

Table 1

Population, Organism, and Tissue Effects Found in Great Lakes Animals

Species	*Population decline*	*Reproductive effects*	*Eggshell thinning*	*"Wasting"*	*Gross defects*	*Tumors*	*Target organ*	*Immune suppression*	*Behavioral changes*	*Generational effects*
Bald eagle	x	x	x	x						x
Beluga whale	x		n/a		x	x	x	x		
Black-crowned night heron	x	x		x	x					
Caspian tern	x	x		x	x		x		x	x
Chinook/coho salmon	n/a	x	n/a			x	x			
Common tern	x			x			x	x	x	
Double-crested cormorant	x	x	x	x	x		x		x	x
Forster's tern	x	x	x	x	x		x		x	x
Herring gull	x	x	x	x	x		x	x	x	x
Lake trout	x	x	n/a	x					x	x
Mink	x	x	n/a	x			x			
Osprey	x	x	x							
Otter	x		n/a							
Ring-billed gull	x		x	x			x			
Snapping turtle	x	x		x	x		x			x

Source: Colborn et al. (1990).

x, Observed effects that have been reported in the literature. Cells with no entry do not necessarily mean there is no effect, only that no citation was found.

n/a, Not applicable.

2 species. Colborn believed that her matrix had yielded a curious result: "Health anomalies have been found in wildlife populations holding significantly higher concentrations of chemicals than unaffected populations. The majority of effects are found in the offspring of wildlife; parental exposure seems to be the cause of problems that are more developmental than carcinogenic."[29]

Her study of the wildlife literature completed, Colborn observed that the animals around the Great Lakes were, for the most part, not dying of cancer. Moreover, whereas the adult animals appeared unharmed by the pollutants, some of their offspring were not surviving, and those that did were afflicted with a variety of abnormalities of reproduction, metabolism, thyroid function, and sexual development. "Everything driven by the endocrine system seemed to be affected."[30] By 1989 Colborn had begun to fit the pieces of the wildlife puzzle together in the shape of the environmental endocrine hypothesis. Two associates who worked with her on the International Joint Commission (a binational agency that oversees the Great Lakes) described her discovery as an "epiphany."[31] The period of discovery is described in *Our Stolen Future:* "Colborn began entering the findings from the studies on a huge ledger sheet, the kind used by accountants. When that became unwieldy, she turned to her computer and created an electronic spreadsheet. . . . As she made entries under columns headed 'population decline,' 'reproductive effects,' 'tumors,' 'wasting,' 'immune suppression,' and 'behavioral changes,' her attention came to focus increasingly on fifteen of the forty-three Great Lakes species that seemed to be having the greatest array of problems."[32]

She reached two conclusions. First, the animals she identified were top predators that consumed Great Lakes fish. Thus, contaminants were more highly concentrated in these organisms, as expected through biomagnification. Second, the health problems were found primarily with the offspring and not the adults. "Now the pieces were beginning to fall together. If the chemicals found in the parents' bodies were to blame, they were acting as hand-me-down poisons, passed down from one generation to the next, that victimized the unborn and the very young. The conclusion was chilling."[33]

Great Lakes, Great Legacy? advanced two unconventional ideas regarding chemical impacts on wildlife. First, the authors stated that chemical exposure affecting reproduction and development might be more important to the survival of wildlife populations in that region than cancer or toxic poisoning. Second, they emphasized that there was evidence of transgenerational effects in which "some stress on the parents . . . affects the well-being of an embryo or newborn."[34] A parent may be hindered in its ability to care properly for its young because of chemical exposure, or the normal development of the offspring may be disrupted owing to exposure either while being carried by the mother, within the egg, or shortly after birth. Within this notion that nonlethal exposure of the parents could severely compromise the survival, the development, and perhaps even the eventual reproductive functioning of the next generation lay the seeds of what would become the environmental endocrine hypothesis.

The global ramifications of the inchoate hypothesis of sublethal exposure and generational effects first hinted at in *Great Lakes, Great Legacy?* were unappreciated at the time. Nowhere in the book do the authors explicitly link the reproductive problems of wildlife to a general phenomenon of endocrine system dysfunction, although a brief section describes "hormonal changes." Only in retrospect can one see the threads of the new hypothesis emerging.

Colborn's literature search for the Great Lakes study spanned the fields of biology, biochemistry, toxicology, and zoology.[35] She discovered evidence of a variety of effects, including incomplete sexual development of male fish in the Great Lakes, increased sterility among wildlife, a decline of chick survival among birds, and thyroid abnormalities among wildlife in the Great Lakes. From these observations she advanced a generalized causal hypothesis, suggesting that the pervasive presence of chemical endocrine disrupters introduced into the environment was responsible for the reproductive abnormalities found in wildlife populations. At the time, Colborn was dubious that anyone would believe her theory. "It's like a science fiction story, . . . to think that we have released thousands of chemicals into the environment and know very little about any of them and some of them are turning out to be capable of actually directing the development of our embryos."[36]

No sooner had the Great Lakes wildlife study been completed than Canada's minister of health asked Health and Welfare Canada to undertake a study of the human effects of the Great Lakes chemicals, and Health and Welfare Canada contracted with Colborn, by then a senior fellow at the Conservation Foundation, to research the human health effects.[37] She identified a group of chemicals that were found in wildlife and also in human tissue, and she produced another matrix of 17 chemicals that had been identified in the Great Lakes study. She learned that each of these chemicals was a developmental toxicant that affected the endocrine system in some way.

Colborn's work eventually caught the attention of another individual with interests in wildlife protection who was in a position to provide her with a supportive institutional context for pursuing her investigations. John Peterson ("Pete") Myers had first become aware of the role that chemicals play in interfering with sexual development in 1975 when he was a graduate student in zoology at the University of California at Berkeley. Myers became acquainted with the discoveries of George and Molly Hunt of the University of California at Irvine, who had observed female-female pairings of gulls, an unusual behavior for this avian species. He also knew of Michael Fry's research on the effect of synthetic pesticides on the sexual development of birds. Fry, a wildlife toxicologist at the University of California at Davis, had exposed eggs from uncontaminated sites to organochlorines and documented abnormalities in the sexual development of the hatchlings that paralleled observations in areas heavily sprayed with pesticides. These reports made a deep impression on Myers, who had been working at the Bodega Marine Laboratory during his graduate studies.[38] Myers had devoted his career to investigating the conservation biology and population ecology of migratory birds. After receiving his doctorate, he continued to pursue his interests in chemical-wildlife interactions, particularly the effect of pesticides on the migratory orientation of birds.

In the mid-1980s Myers became the vice president of research at the National Audubon Society. He gave a series of lectures in which he drew attention to unresolved concerns about the chemical factors influencing migratory bird patterns. At one of these lectures in 1988, which took place at the Washington, D.C., headquarters of the Audubon Naturalist Club, a local independent conservation organization, Theo Colborn was in the audience. She heard Myers discuss the disappearance of migratory shorebirds, which he attributed to aggressive pesticide use. She took the opportunity to express her enthusiasm for his talk and to share with him her findings from the Great Lakes study. Myers in turn invited Colborn to give a seminar to his staff at the National Audubon Society. He considered her work of sufficient promise that he wanted to hire her, but the negotiations with the society did not work out. Coincidentally, during that period Myers received an offer from the W. Alton Jones Foundation to serve as its director. He used the negotiations over his appointment to interest the foundation in creating a senior fellows program.

Soon after accepting the position at W. Alton Jones, Myers brought Colborn on in 1990 as one of the foundation's first two senior fellows. For three years she had the opportunity to pursue her ideas about chemically induced effects on wildlife without having to think about her salary or raising money. Between 1989 and 1990 Colborn wrote a paper for the National Academy of Sciences in which she stated: "These [endocrine-disrupting] contaminants may have more control over the behavioral development of offspring of exposed adults than the genes the offspring inherit, or the training they receive."[39] She recognized that the hypothesis she was advancing represented a challenge to the tenets of traditional toxicology, epidemiology, and chemical risk assessment, particularly in how they approached low-dose exposures of endocrine disrupters.

By the early 1990s, with continued support from W. Alton Jones, Colborn was building the elements of a new paradigm to understand the role that chemicals play in wildlife decline. This included a search for a more encompassing end point she described as "survival of the species." For Colborn the next step in the advancement of the hypothesis was to share her findings with scientists from other disciplines and to have them reexamine their specialized fields of knowledge in light of the endocrine disrupter evidence. One of the scientists she had contacted soon after formulating her conjecture was Frederick vom Saal, a reproductive physiologist from the University of Missouri who studied the effect on animal behavior and physiology of small hormonal changes during fetal development. Trained as a neurobiologist at Rutgers University, vom Saal had focused during his postdoctoral appointment at the University of Texas and later at the University of Missouri on critical windows in reproductive development. He investigated the extent to which chemicals that determine which genes get turned on or off in cells have a lasting effect on physiology and behavior.

Vom Saal published two seminal articles in *Science* in which he studied litter-bearing pregnant mice.[40] He carried out cesarean sections on the animals and discovered a phenomenon called the *positioning effect.* During delivery, male mice positioned in the womb between two females receive an extra dose of estradiol (natural estrogen) that diffuses through the amniotic fluid. The additional estrogen exposure influences the behavior and physiology of the male offspring. A similar testosterone effect is found when sister mice are situated in the womb between two males. The differences in hormone exposure caused by the positioning of the mice in the uterus were quite small, yet the behavioral and physiological effects were nonetheless significant. Some animals that experienced this positioning effect became more aggressive and territorial—the result of one-time exposures to additional estrogen (or testosterone) that seemed to have made imprints in their brains. These experiments revealed that even minute changes in the hormone exposure of the developing fetus during certain sensitive stages could result in measurable effects. Could the positioning effect be replicated by xenoestrogens?

In the fall of 1989, Colborn contacted vom Saal and informed him that she believed his work on animal endocrine effects was relevant to the damage wildlife had sustained from environmental chemicals. Although vom Saal was aware of some of the research on the impacts of organic chemicals (DDT and methoxychlor) on animal reproductive physiology, he had never related it to his own work because the former studies involved exposures to concentrations that were orders of magnitude higher than those of the estrogenic substances he was investigating.

Colborn showed vom Saal extensive documentation on what she had discovered about endocrine disrupters in wildlife. After spending a couple of weeks reviewing her materials, vom Saal concluded that she was onto something big—a major new hypothesis. By 1989 Colborn had a clear picture that chemicals released into the environment were interfering with the role of endogenous hormones and adversely affecting the healthy development of certain organisms.[41]

Vom Saal began redirecting his research focus in 1990 as a result of his contact with Colborn.[42] He questioned whether the endocrine effects of xenobiotics fit into the standard paradigm of toxicology, and if they did not, he wondered how the standard paradigm would have to be modified to account for the role of chemicals as hormone disrupters. Traditional toxicologists were generally ill informed about the dose-response effects of hormones. Endocrine systems were homeostatic (based on self-regulating feedback systems), and thus standard monotonic dose-response curves (the higher the dose, the greater the response) did not, as a general rule, apply.

By reaching out to vom Saal and other scientists, Colborn began a process of organizing interdisciplinary work sessions that enabled scientists to build bridges across fields, to compare notes about different species, and to learn about how their own work informed and was informed by the environmental endocrine hypothesis. Vom Saal was able to put her in touch with leading reproductive physiologists as well as clinical and animal endocrinologists. Colborn herself was already well networked in wildlife biology circles. The result was a unique cross-fertilization of scientific disciplines that widened the circle of interest in endocrine disrupters.

Meanwhile, Pete Myers used his position at W. Alton Jones to introduce the issues of endocrine disrupters to three new constituencies: environmental activists concerned about the effects of toxins on human health and wildlife, the foundation community, and public health physicians and pediatricians. While Colborn was networking with scientists, Myers worked through his three constituencies to build a broader public awareness of the wildlife evidence for and the potential human consequences of endocrine disrupters. He later noted: "My impact was the implicit encouragement I gave [to the constituencies] to think about the issue Theo was raising."[43] As the director of a major foundation that awarded between $13 and $20 million in grants annually during the 1990s, Myers met with many potential grant recipients. These individuals brought their own ideas to the foundation, but they also paid close attention to the foundation's new initiatives. Eventually, groups began to incorporate concerns about endocrine disrupters into their organizations' goals. During the period 1991–92, W. Alton Jones awarded $80,000 for research and $234,000 for advocacy, organizing, and public education on pesticides and endocrine disrupters. In 1993, the foundation formally introduced the program category "endocrine disrupters and pesticides" and funded five projects amounting to $450,000. In subsequent years the grants awarded under this category grew to over $1 million.

Some of the organizations that began to consider endocrine disrupters in their work included the Natural Resources Defense Fund, Environmental Media Services, the Environmental Working Group, Physicians for Social Responsibility, and the Children's Environmental Health Network. These and other nonprofit advocacy organizations spread the message about endocrine disrupters in the context of their overall programs directed at reducing levels of industrial organochlorines and pesticides in the environment.

Myers reached out to other foundation directors when he served on the board of the Consultative Group for Biological Diversity, which included representatives from some 30 foundations. In 1994, he introduced Colborn and her *Our Stolen Future* coauthor Dianne Dumanoski to the Consultative Group. Eventually, some of these foundations began funding organizations that communicated the issues to the public and the media. The growth of scientific and public awareness of hormone-disrupting chemicals was occurring along parallel but mutually reinforcing paths.

The First Wingspread Work Session

The first in a series of meetings convened by Colborn, with the help of Myers and the support of the W. Alton Jones Foundation,* took place in July 1991 and was titled "Chemically Induced Alterations in Sexual and

*Other support came from the Charles Stuart Mott Foundation, the Joyce Foundation, the Keland Endowment Fund of the Johnson Foundation, and the World Wildlife Fund.

Functional Development: The Wildlife/Human Connection" (see Appendixes A and B). Known as the Wingspread Work Session or simply Wingspread I (it was held at the Wingspread Conference Center in Racine, Wisconsin), it brought together participants who were sufficiently concerned about the significance of what they had learned from one another that they issued a consensus statement citing the potential public health implications of their findings.

For those who have followed and participated in the development of the environmental endocrine hypothesis, Wingspread is described as a watershed event—one that expanded both their vision and the potential scope of the problem of chemical effects. Frederick vom Saal, one of the participants, characterized it as follows:

> Wingspread was like a religious experience. We got together in working groups after the presentations. The evidence presented was incredible. Theo was a taskmaster working 20 hours a day. People realized the monumental information that was being put together there. The weekend put things into perspective for me. The magnitude and seriousness of the problem became very clear. People were stunned by what we learned there. It doesn't happen very often that you see things totally differently after a short period of time. It was a turning point.[44]

Up until this period, it was John McLachlan at NIEHS who had been the guiding force behind interdisciplinary discussions of environmental estrogens. The meetings McLachlan facilitated dealt largely with DES, its human effects, and a few chemicals that were analogous in function and action. The first "Estrogens in the Environment" conference, organized by McLachlan in 1979, included presentations on the estrogenicity of kepone in birds and mammals, DDT, and mycotoxins.[45] In 1987, before Colborn had publicly advanced her theory, McLachlan coauthored a paper in *Environmental Health Perspectives* in which he raised the possibility that chemicals commonly found in the environment may exhibit estrogenic properties: "It is clear that exposure to exogenous estrogens during fetal development will profoundly alter sexual differentiation. . . . These findings with the potent synthetic estrogen, DES, provide a possibly exaggerated model for what might be seen with much weaker estrogenic xenobiotics, many of which are common features in our environment."[46]

McLachlan identified DDT, kepone, and certain polycyclic aromatic hydrocarbons (e.g., benz[*a*]anthracene) as estrogenic. He also cited natural estrogenic substances (phytoestrogens), such as clover, that induced animal infertility. He dubbed these the "new category of environmental compounds, hormonally active xenobiotics."[47]

While McLachlan was helping lay the scientific foundations for the study of xenoestrogens, he recognized Colborn's unique contribution to environmental toxicology and in particular—through her linking of the wildlife and human effects of estrogenic chemicals—her extension of his work.[48] Referring to Wingspread I, he noted: "Theo pulled us together. . . . We hadn't read each other's literature. Her main contribution is linking wildlife and human effects [of estrogenic chemicals]. It was one of her real strokes of cleverness and insight."[49]

The 1991 Wingspread meeting provided a broad panorama of suggestive evidence pointing to a general worldwide phenomenon linking chemical exposure to the alteration of normal sexual development. The wildlife data presented at Wingspread I afforded some of the most compelling evidence for a connection between persistent chemicals in the environment and destabilizing reproductive effects mediated by disturbances of the endocrine system. But the ultimate power of the wildlife data in gaining attention for the hypothesis was that they could be interpreted as providing an "early warning system" for human difficulties, as had earlier been argued in *Great Lakes, Great Legacy?*

Evidence from widely disparate locations pointed to reproductive problems in freshwater, marine, and terrestrial organisms exposed to xenobiotics.[50] These environmental chemicals could be shown to exhibit biological activity similar to that of endogenous hormones or to inhibit hormonal action in controlled experiments. Thus their presence, especially during critical periods in embryonic growth, could cause disruption of the normal processes controlling development. The wildlife evidence demonstrated that although parental animals appeared healthy, their offspring exhibited abnormal development and disruption of their physiological systems. Exposure to xenobiotics was related to the early death of embryos, loss of fertility, behavioral changes, and morphological and physiological changes in the reproductive system and other organ systems that were endocrine targets.

The Wingspread meeting was notable on two counts. First, the meeting brought together as diverse a representation of scientific fields as has ever been convened on the subject of endocrine disrupters—including researchers in human and wildlife toxicology, endocrinology, anthropology, ecology, immunology, histopathology, and wildlife management—who shared their methodologies and data on the issue of abnormal sexual development. Although conferences on the human effects of some of these chemicals had been taking place since the late 1970s, this was the first time that wildlife biologists and human health scientists had had an opportunity to talk with one another. Colborn's Great Lakes study provided the basic insight that it was unwise to focus on cancer and acute toxicity to the exclusion of other end points in studying the effects of chemical contamination. The Wingspread meeting reinforced and deepened earlier conjectures about developmental effects.

A second significant achievement of Wingspread was that Colborn and Myers guided the scientists to a consensus statement that drew attention to the serious environmental consequences of their findings. The idea behind the consensus statement and the form it took came from Myers, who drew from his experience with the consensus process of scientists involved in the global warming debates. Myers recalled: "I brought in the formula about the structure for reaching a consensus statement. I forced the consensus discussion to that forum and argued why it was right as a structure, why it was important. I was the facilitator of that consensus; Theo was the architect."[51] In the scientific assessment of global warming prepared by the Intergovernmental Panel on Climate Change, the conclusions reached and supported by signatory scientists were scaled according to confidence level and distinguished in certitude by phrases such as "We are certain of the following," "We calculate with confidence," or "Based on current model results, we predict . . ."[52]

The four-page Wingspread consensus statement provided the first detailed formulation of the hypothesis and offered a unifying thematic focus for the diverse research agendas. It established a baseline of knowledge on which future meetings could build, and it gave scientists an opportunity to make some policy pronouncements, such as that testing programs for chemicals must be expanded, that ecosystem regulations must be put into place, and that the cancer paradigm for evaluating the effects of chemicals in humans and wildlife was wanting. Wildlife studies were central to the development of the hypothesis, with the participants agreeing that populations of animals are being harmed by endocrine disrupters. However, no one was naive enough to believe that the importance of the hypothesis could be established without understanding its implications for humans. Many of the wildlife studies were based on animals exposed to unusually high levels of endocrine disrupters. It remained unclear whether these data had relevance to human risk, since human exposures to this class of chemicals were at much lower doses and as yet not well understood.

The importance of the Wingspread I consensus statement, signed by 21 scientists, was symbolically noted in the published proceedings volume by its placement within the first eight pages as the prolegomenon to the scientific papers.[53] Following the model set by the global warming consensus position, the Wingspread statement was divided into section headings signifying confidence levels graded from higher to lower: "We are certain of the following," "We estimate with confidence that," "Current models predict that" (including a section on model uncertainties), and "Our judgment is that. . . ." The statement ended with a discussion of ideas for improving predictive capability.

The organization of the statement into different levels of confidence showed a deep understanding of the scientific ethos. Scientists are likely to understate their data when asked about causal connections. In the scientific culture, one learns about the importance of issuing caveats before any claims for causality are made. Colborn and Myers greatly increased the likelihood of gaining full support for the statement by establishing gradations of confidence, thus building in sufficient caveats on cause and effect. In the subsection of the consensus statement reflecting the highest certitude ("We are certain of the following") the caveats are reflected in the use of the terms "potential" and "may":

> We are certain of the following.
>
> —A large number of man-made chemicals that have been released into the environment, as well as a few natural ones, have the potential to disrupt the endocrine system of animals, including humans. Among these are the persistent, bioaccumulative, organohalogen compounds that include some pesticides (fungicides, herbicides, and insecticides) and industrial chemicals, other synthetic products, and some metals.
>
> —Many wildlife populations are already affected by these compounds. The impacts include thyroid dysfunction in birds and fish; decreased fertility in birds, fish, shellfish, and mammals; decreased hatching success in birds, fish, and turtles; gross birth deformities in birds, fish, and turtles; metabolic abnormalities in birds, fish, and mammals; behavioral abnormalities in birds; demasculinization and feminization of male fish, birds, and mammals; defeminization and masculinization of female fish and birds; and compromised immune systems in birds and mammals.
>
> —The patterns of effects vary among species and among compounds. Four general points can nonetheless be made: (1) the chemicals of concern may have entirely different effects on the embryo, fetus, or perinatal organism than on the adult; (2) the effects

are most often manifested in offspring, not in the exposed parent; (3) the timing of exposure in the developing organism is crucial in determining its character and future potential; and (4) although critical exposure occurs during embryonic development, obvious manifestations may not occur until maturity.

— . . . The effects seen in in utero DES-exposed humans parallel those found in contaminated wildlife and laboratory animals, suggesting that humans may be at risk to the same environmental hazards in wildlife.

The Wingspread I consensus position had taken a significant leap forward, both in the breadth of its hypothesis connecting wildlife abnormalities to human risks and in the confidence of the assertion that this was a real effect. Not only did the consensus statement bring cohesion to the workshop participants, but it provided a highly effective tool for raising the issues before a nonscientific audience. Less than three months after the Wingspread I meeting, the consensus position was reported to a Senate committee investigating reproductive hazards (see Chapter 2). The consensus statement was subsequently cited in many journalistic accounts and brandished by activists as a new call to arms against environmental contaminants.

With the success of the Wingspread I meeting, the climate was now receptive for investigating other connecting links between synthetic chemicals and reproductive abnormalities. Meanwhile, the relationship between xenobiotic endocrine disrupters and abnormal sperm was being pursued quite independently in Europe. The postulated link between chemicals and sperm count provided yet another lattice in the expanding structure of the theory of endocrine-disrupting chemicals. The social and ethical issues thus raised were to trigger a cascade of media reports linking endocrine disrupters to the "survival of the species."

Environmental Effects on Human Sperm

If the DES story represents one path to the generalized environmental endocrine hypothesis and investigations into wildlife effects a second, then the third path originated in studies of male infertility and testicular cancer. Growth in the use of artificial insemination and in vitro fertilization for conception required methods for assessing sperm health and viability. Commercial sperm banks maintained records of healthy donors. Fertility clinics also began recording sperm data from clients who sought help for infertility. Thus data on the quality and density of human sperm have been collected for about 50 years, albeit in uncoordinated individual projects that sometimes utilized different methods for counting and evaluating the quality of the sperm. But not until recently have scientists systematically analyzed the quality and density of human sperm and its relation to environmental toxicants. With the rapid expansion of sperm banks and fertility clinics, scientists finally had access to data that allowed them to investigate the relationship of sperm characteristics to the age and health status of sperm donors. One of the leading contributors to this area of research is the Danish clinician-scientist Niels Skakkebaek.

Skakkebaek, a pediatrician at the National University Hospital in Copenhagen, heads the Department of Growth and Reproduction, which specializes in male infertility and pediatric endocrinology. After receiving his M.D. degree in the late 1960s, Skakkebaek trained under some of the world's leading cytogeneticists in North America and Europe, while completing his doctorate of medicine and his clinical training. His early research centered on spermatogenesis in infertile males. He chose a specialty in pediatric endocrinology because he believed that reproductive disorders in adult males originated much earlier, possibly in childhood or even during fetal development.

In the early 1970s Skakkebaek discovered an usual group of cells in the testes of men diagnosed with testicular cancer. The cells had the characteristics of fetal cells. It was on the basis of such observations that he conjectured that testicular cancer in young adulthood originates in cells that have already been formed during fetal development. These cells, once formed, remain in the testes and begin to proliferate after puberty when they are exposed to the hormones of adolescence. That, Skakkebaek noted, is the age when tumors in the testes are first observed.

When Skakkebaek published his fetal cell hypothesis in 1972, he was met with skepticism from American pathologists who overlooked the significance of the fetal-like adult cells. It took over 10 years before Skakkebaek's finding of the abnormal testicular cells and his hypothesis that they are linked to testicular cancer were finally accepted by the scientific community.

Skakkebaek and his colleagues were all too aware that Denmark had the highest incidence of testicular cancer in the world. The lifetime risk factor was about 1 percent for all Danish men. By 1980 his research group had begun to ask questions about why the incidence of testicular cancer was so elevated. He learned from Danish epidemiologists that the demography of testicular cancer had changed. At first, the rate of testicular cancer was lower outside Copenhagen than in the city, but subsequently the incidence of the disease began to rise in the suburbs. This fact, and the observation that the increases occurred suddenly rather than gradually over a 40-year period, suggested to him that environmental factors were at work.[54]

To investigate the relationship between abnormal cells of the testis and reproductive disorders, in the mid-1980s Skakkebaek established a sperm laboratory in his research center. He began to find associations between poor semen quality and testicular cancer, cryptorchidism (undescended testes), and intersex conditions. Until he could undertake a study of the sperm quality of "normal" subjects, however, Skakkebaek would have no baseline for interpreting the incidence of reproductive disorders. Through a physician friend employed by an airline, he assembled a reference group of 141 males among the company's ground personnel. His research team discovered that 50 percent of the males in this group had abnormal forms of sperm. This finding seemed quite high and unexpected for otherwise normal subjects. It was of particular concern since abnormal sperm may portend future disease.

At the time Skakkebaek was in contact with the Human Reproductive Program of the World Health Organization (WHO). After the end of the cold war, the deplorable degree of chemical pollution in eastern bloc countries had become widely known, and WHO turned its attention to the hazards that chemical toxins posed to reproduction in eastern Europe. Along with six other groups—including the Danish Ministry of the Environment, the Commission of the European Communities, and the NIEHS in the United States—WHO sponsored a 1991 workshop on the environment and reproduction two months after Colborn had organized the first Wingspread Work Session on endocrine disrupters. Skakkebaek was made chairman of the steering committee. The proceedings of the meeting were published in a special supplement to *Environmental Health Perspectives*.[55]

The workshop at last brought together McLachlan and Skakkebaek, each of whom contributed an important piece to the general hypothesis regarding the mechanism of xenobiotic endocrine disruption. This was their first opportunity to become acquainted with each other's work. McLachlan, who had also attended Wingspread, served as a bridge between the European and American approaches to xenoestrogens and between human reproductive and wildlife issues.

At the WHO meeting Skakkebaek delivered a talk on sperm decline, and a British newspaper covered the presentation.[56] This was to be the start of Skakkebaek's protracted and disconcerting relationship with the British press—and eventually the international media as a whole.

On recommendations from his colleagues, Skakkebaek submitted his study of sperm in "normal" subjects for publication, and in 1992 his results appeared in the internationally renowned *British Medical Journal.* The article was devoted primarily to an analysis of data on sperm decline, but in the conclusion the authors stated: "Such remarkable change in semen quality and the occurrence of genitourinary abnormalities over a relatively short period is more probably due to environmental rather than genetic factors. . . . Whether oestrogens or compounds with oestrogen-like activity . . . or other environmental or endogenous factors damage testicular function remains to be determined."[57]

No single study included enough human subjects to justify a generalized finding of global sperm decline. How widespread was the phenomenon? To answer this question, Skakkebaek and his collaborators undertook a meta-analysis of 61 papers published between 1938 and 1990 involving nearly 15,000 men. A meta-analysis combines various types of summary statistics from independent studies into an overall statistic for the group of studies—in effect, it is a study of studies. Skakkebaek and his colleagues reported a significant decrease in mean sperm count, from 113 million per milliliter in 1940 to 66 million per milliliter in 1990, and concluded there was strong evidence that the density (quantity per milliliter) of human semen had decreased steadily over the past 50 years. The studies on which they based their results covered 21 countries in North and South America and Asia. Most of the studies confirmed the thesis that human sperm was deteriorating in both density and quality. Finland was one of the exceptions. Not only had Finland reported no change in sperm concentration between 1958 and 1992, it also reported sperm density values higher than those elsewhere in Europe.[58]

The article drew extensive media coverage—not all of it accurate. Skakkebaek was especially chagrined by the ludicrous interpretations placed on the results, such as the extrapolation of the linear decline of sperm counts to zero in 30 years, resulting in absolute infertility.

Studies of the sons of women given DES and workers exposed to the pesticide dibromochloropropane (DBCP) suggested to some scientists that there was an environmental cause behind the declining sperm counts. However, until 1993 there were no published hypotheses in the major journals seeking to explain the phenomenon.

A year after publication of the *British Medical Journal* article Skakkebaek collaborated with Richard Sharpe, a reproductive biologist in Edinburgh, Scotland, whom he had known from visits to Edinburgh. The two scientists coauthored a paper in the British medical journal *The Lancet* that postulated a link between xenoestrogens and sperm count decline.[59] The paper had first been submitted to *Science,* but it was returned without having been sent out for review. Known for its high rejection rate, *Science,* while not yet ready to publish papers on environmental endocrine disrupters, was nevertheless willing to discuss the issue in its news section.

The title of the Sharpe-Skakkebaek paper took the form of a question: "Are Oestrogens Involved in Falling Sperm Counts and Disorders of the Male Reproductive Tract?"[60] Even though the authors were careful to frame their hypothesis as an interrogative, public interest groups were quick to cite it as scientific support for their stand against synthetic chemicals. Greenpeace published advertisements in Sweden that raised concerns about estrogenic chemicals and, according to the authors, misused their results to support the organization's agenda.[61] In response, the authors issued a press release disassociating themselves from Greenpeace's interpretation of their results. After publication of the *Lancet* article, Skakkebaek was contacted by the British Broadcasting Corporation to participate in a documentary broadcast on xenoestrogens and male reproduction, to be titled *Assault on the Male* (see Chapter 2).

Skakkebaek and Sharpe had developed their version of the environmental endocrine hypothesis quite independently of the events taking place in the United States. They were aware of (and in fact cited in their references) the proceedings of the 1991 Wingspread Work Session and McLachlan's 1985 "Estrogens in the Environment II" meeting (see Appendix A). But these citations were part of a general background of supporting or closely associated studies. The authors had already framed and advanced a model explaining the effect of estrogens on development of the testis and reproductive tract, their effect on Sertoli cells (cells involved in sperm production or spermatogenesis), and the relationship between Sertoli cells and spermatogenesis.

Another relevant area of research focused on occupational hazards to fertility. By the 1980s, occupational studies of pesticide workers had revealed that exposure to chemicals could reduce sperm count and impair the quality of sperm.[62] Of particular significance in the occupational health literature were incidents like the reported infertility of pesticide plant workers exposed to DBCP.[63]

Skakkebaek and Sharpe were also influenced by what they had learned about the DES episode. Between 1945 and 1971 as many as 3 million pregnant women in the United States had been treated with DES, exposing between 1 and 1.5 million males in utero to the synthetic estrogen. The otherwise normal adult sons of treated women were reported to have decreased semen volume and sperm counts.[64] The connections Skakkebaek and Sharpe drew between maternal estrogen concentration and male reproductive disorders such as testicular cancer and cryptorchidism prompted them to think about the effect that foreign estrogens might have on the quality and density of sperm. Having referenced the second "Estrogens in the Environment" conference (1985) and the first Wingspread Work Session (1991) in their paper, Sharpe and Skakkebaek were clearly building on the preponderance of new data on endocrine effects reported at those meetings when they wrote: "This possibility [of a connection between xenoestrogens and sperm count] is not unlikely given the view that 'humans now live in an environment that can be viewed as a virtual sea of estrogens.'"[65]

The authors also cited studies in rodents of single maternal neonatal exposures to traces of dioxin and PCBs that demonstrated reduced semen quality in adult offspring. They acknowledged the conjectural nature of the inference drawn between animal and human effects ("It is not known whether exposure of humans to these chemicals is sufficient to induce 'oestrogenic' effects"); although they reported that similar effects had been observed in wildlife.[66]

In their 1993 article, Sharpe and Skakkebaek proposed what they viewed as a plausible mechanism linking overexposure to estrogens or estrogenic chemicals in fetal life to declining sperm counts, sperm quality, and sperm motility. At the heart of the explanatory mechanism were the Sertoli cells. The authors maintained that the number of Sertoli cells is determined in large part by the level of follicle-stimulating hormone (FSH). Inhibition of FSH secretion during certain stages of fetal development, they argued, can result in a reduction of Sertoli cell production. Moreover, Sertoli cell number in early development determines testicular size and sperm output in adulthood. They emphasized that FSH secretion in neonates, which could also influence the production of Sertoli cells, is highly sensitive to inhibition by xenoestrogens. Thus, in brief, the mechanism is as follows: a pregnant woman is exposed to estrogenic chemicals; these chemicals cross the placenta, exposing the developing fetus; FSH secretion is inhibited during gestation; consequently Sertoli cell production is low, eventually affecting adult sperm output and quality.

The Sharpe-Skakkebaek hypothesis began to attract international attention to the root causes of male infertility in ways that the mere reporting of sperm count declines could not. Their conjecture linked the generalized environmental endocrine hypothesis with a measurable global reproductive effect in humans. Nevertheless, evidence for regional declines in human sperm counts had been widely but not universally accepted, and the mechanism behind the effect, and its link to the environmental endocrine hypothesis, was still largely speculative.[67]

Nothing stimulates as much public interest in biomedical science as postulating a cause for an important human disease or abnormality, particularly when that cause supports a new unified theory of disease. By 1995 the conjecture linking xenoestrogens to sperm count density began to receive serious consideration beyond intermittent publications in science journals. Subsequent to a January 1995 workshop on male reproductive abnormalities held in Copenhagen and organized by Skakkebaek, the Danish Environmental Protection Agency issued a report titled *Male Reproductive Health and Environmental Chemicals with Estrogenic Effects,* which highlighted the Sharpe-Skakkebaek hypothesis that adverse effects in male reproductive health observed in wildlife and humans were the result of increasing exposure to estrogenic pollutants in the early stages of development.[68]

With both American and European science focusing on male fertility, it was only a matter of time before the issue hit the mass media as a science exposé. In January 1996, the *New Yorker*—the favored magazine of the literary intelligentsia in the United States, and the one that had serialized *Silent Spring*—published a journalistic essay titled "Silent Sperm."[69] The report offered as many as a dozen hypotheses for the decline in male fertility, including exposure to gasoline fumes, sedentary lifestyle, and stress. But the author gave considerable attention to Sharpe and Skakkebaek's theory that the decline in sperm count was caused by environmental assaults on the reproductive organs, and that it was correlated with testicular cancer, undescended testes, and hypospadia (an abnormal placement of the urethral opening). As the *New Yorker* essay wades through the maze of scientific explanations, the reader is left with uncertainty that there is in fact any single cause. Moreover, the author, paraphrasing Sharpe, states that the hypothesis is still unproven and that, after all is said, the decline in sperm count may be due to environment, lifestyle, or even an evolutionary mutation.

The doubts raised in the *New Yorker* article regarding environmentally induced sperm decline were reinforced by two widely publicized studies published in the *Fertility and Sterility Journal* that reported on sperm counts of men in New York, Los Angeles, Seattle, and Roseville, Minnesota. In one study, investigators examined data from semen analyses of 1,283 men who had donated to sperm banks before having vasectomies;[70] the other was based on data from 500 men who had donated sperm for a clinical study.[71] These studies found that the sperm counts of the men in these cities had not declined over a 20-year period, and furthermore that New York men have among the highest sperm counts in the world. The *New York Times* reported the findings under the headline "Are U.S. men less fertile? Latest research says no."[72] One of the scientists involved in the study called into question the general hypothesis that worldwide sperm levels were down: "I can explain all the decline in sperm counts by geographical variability." These results were used by skeptics to discount the large body of research cited by Skakkebaek and his colleagues as the basis for their assertion that the decline in sperm count was real and that it was probably due to environmental factors.

Yet supporters of the sperm decline theory found strength in numbers. By August 1996, 19 scientists had coauthored a long scientific review titled "Male Reproductive Health and Environmental Xenoestrogens."[73] Published in *Environmental Health Perspectives,* the journal most sympathetic to the environmental endocrine hypothesis, the review was a revised version of the report on male reproductive problems and xenoestrogens issued by the Danish Environmental Protection Agency one year earlier.[74] Although it appeared in print four months after the publication of three articles critical of the sperm decline theory, there were no references to these new studies. The multiauthored review highlighted many concerns about the declining reproductive health of men. On the issue of sperm counts the authors concluded that it was "now evident that several aspects of male reproductive health have changed dramatically for the worse over 30 to 50 years. The most fundamental change has been the striking decline in sperm counts in the ejaculate of normal men. . . . The result is that many otherwise normal men now have sperm counts so low that their fertility is likely to be impaired."[75]

Six years after Skakkebaek and Sharpe had published their hypothesis that a worldwide decline in sperm quality and density was being caused by chemical exposure, scientific debate focused on whether reports of such a decline were truly representative of the actual trends. Many of the studies used in the authors' meta-analysis were vulnerable on grounds that there had been no standardized procedures in place for collecting, sampling, or measuring sperm. The sperm donors, their lifestyle, their health, and their motivation for participating were all relevant to the outcome. A meta-analysis of 61 studies simply compounds errors or bias that may be found in each of the underlying studies. Some scientists, while debunking the idea that sperm decline was a universal trend, accepted the possibility of regional declines, citing certain industrial regions that did not follow a downward trend. The debate began to shift away from the raw data to the statistical modeling techniques that one must use to interpret the data in the 61 studies. Epidemiological statisticians sparred over the uses of linear versus nonlinear regression models (mathematical expressions that are best fits to the observed data when plotted) to determine whether the data points for sperm quantity were indeed on the decline. These debates prompted nonstatisticians to question whether sperm decline was simply an artifact of modeling—a social construct—rather than an objective fact.

Shanna Swan and her colleagues at the California Department of Health Services used several models in their own reanalysis of the 61 studies and concluded that, regardless of whether a linear or a nonlinear model was used to interpret the data points, the results were basically the same. They also checked the data for confounding factors—such as differences in sperm counting methods or sexual abstinence time of the sperm donors—that might have explained the decline. According to their findings, "The decline in sperm density reported by Carlsen et al. is not likely to be an artifact of bias, confounding, or statistical analysis"; they concluded that the data showed a significant decline in sperm density in the United States and northern Europe but not in non-Western countries.[76]

Two points should be made about the debate over sperm counts. First, studies that indicate robust sperm levels in specific cities or regions are not necessarily inconsistent with overall global trends of declining sperm levels. Individual exceptions do not invalidate trend data. As an example, a year of lower-than-average global temperatures does not disprove a trend toward global warming. Second, if the conclusion that there is a declining trend in sperm quality and quantity for Europe and the United States is validated in future studies, elucidating the cause(s) of that trend remains a formidable problem.

Despite the uncertainties over the causes of declining male fecundity, proponents of the endocrine hypothesis consider reports linking sperm decline with chemical endocrine disrupters to make up a growing part of the inconclusive yet plausible causal stories that, in their aggregate, strengthen the theory. The discovery of declining sperm counts has been linked to other abnormalities of the male reproductive system, and so the generalized hypothesis provides a framework and added justification for examining more thoroughly other possible outcomes of exposure to endocrine disrupters, such as testicular cancer.[77] While theories on the genetic basis of cancer are gaining hegemony in the scientific world, the environmental endocrine hypothesis provides one of the most important new paradigms for environmentally induced disease. The range of the hypothesis also encompasses a postulated mechanism of cancer causation. Ironically, the wildlife data that Colborn compiled offered very little evidence linking chemicals to cancer, and she eventually turned her attention to new end points of disease. But the link between hormones and cancer, as illustrated by the DES episode, was not new, and it provided another way of thinking about endocrine disrupters.

Breast Cancer and Estrogenic Chemicals

Breast cancer is the most commonly diagnosed cancer among women and the second leading cause of cancer-related death among women after lung cancer. Since 1973 there has been a significant rise in the number of diagnosed cases of breast cancer in women older than 50. The rate for all women has risen slightly. The incidence of breast cancer in women younger than 50 has held steady. Between 1973 and 1989 the incidence of breast cancer increased by 21 percent, about 1 percent annually. In the 1940s, the rate of breast cancer was about 58 cases per 100,000 people; in 1990 it was over 100 per 100,000. The lifetime risk of breast cancer has climbed from 1 in 20 at the end of World War II to 1 in 8 by the mid-1990s—more than doubling the risk.

The steady increase in the incidence of breast cancer and the lack of a satisfactory explanation inspired the mobilization of local and regional breast cancer activists and women's support groups to form the National Breast Cancer Coalition in 1991. Federal spending on breast cancer research in the United States rose from about $90 million in 1990 to over $600 million in 1999, testifying to the importance of a national constituency and aggressive lobbying efforts in addressing the problem.[78] Other regional activist groups, like One in Nine, which was established on Long Island in response to the high breast cancer rates in Nassau County, lobbied their congressional delegations for new legislation targeting cancer-causing chemicals (see Chapter 2).

The relationship between xenobiotic chemicals and breast cancer was suggested relatively late in the formulation of the environmental endocrine hypothesis. Scientists have been hesitant about making the association between environmental endocrine disrupters and the risks of breast cancer, despite the fact that the connection between breast cancer and hormones has been established for some time.[79] A British surgeon, writing in *The Lancet* in 1896, reported the marked improvement of two women with advanced breast cancer whose ovaries had been removed surgically (a procedure called an oophorectomy). By the early 1930s scientists had demonstrated experimentally that certain types of estrogen, when injected into mice, could induce breast cancer, establishing a sufficient if not a necessary causative connection between exogenous estrogens and cancer for those animals at the specified doses.

For several decades androgen therapy was used to treat human breast cancer because of observations that the growth of mammary carcinoma in mice could be inhibited by the administration of testosterone. In the mid-1940s estrogen therapy began to compete with androgen therapy as the preferred treatment for breast cancer. In 1969 Stoll reported that "oestrogen therapy given as a primary therapy will induce tumour regression in over 30 percent of postmenopausal patients with advanced breast cancer."[80] So while estrogens were identified as part of the etiology of cancer, they were also called upon as a therapeutic solution. By the 1970s science had come to believe that breast cancer was "hormone responsive," meaning that endocrine modification could lead to stimulation or regression of breast tumors.

Clinical and epidemiological studies confirmed the connection between certain forms of estrogen exposure and cancer. For example, there was the discovery that DES caused clear cell adenocarcinoma of the vagina, cervix, or both in the offspring of women treated with the synthetic hormone. In addition, surveys of women have revealed breast cancer risk factors that include lower age of menarche and late menopause. In general, women with breast cancer have higher levels of serum estrogen than women without breast cancer. These experimental data prompted observations such as the following: "Women in Western nations like our own are over-estrogenized. They produce more of the hormone than is needed for normal reproduction function. Oriental [*sic*] women have lower amounts of estrogen in their bodies. . . . The excess estrogen in American women and other Western women thus may serve no useful purpose—but does seem to promote breast cancer."[81]

So while many reports in the 1980s began postulating total estrogen load, including high in utero estrogen levels, as a risk factor in breast cancer, researchers focused on food and lifestyle as the primary sources of "over-estrogenation." The possibility that ambient chemical exposure might produce sufficient levels of estrogen mimics to induce an adverse biological effect seemed remote. There were studies that suggested a possible link between xenoestrogens and breast cancer, but no major publication in a peer-reviewed journal posited the connection and a possible mechanism for it until 1993.

Years before the generalized hypothesis of endocrine disrupters was advanced by Colborn, Leon Bradlow, professor of biochemistry in surgery and pediatric endocrinology at the Cornell University Medical School, was investigating estradiol metabolism in the context of breast cancer. Bradlow was chief editor of the journal *Steroids* from 1985 to 1993. He began studying the metabolism of estradiol in 1977, and by 1980 he had published two articles on the effects of estradiol on men and women. In 1981 he reported the increase of a particular estradiol metabolite (16α-hydroxyestrone) in systemic lupus, and a year later observed abnormal oxidative metabolism of estradiol in women with breast cancer.[82] He found a 50 percent elevation of 16α-hydroxyestrone in women with breast cancer, a significant elevation in women at risk for breast cancer, and elevated amounts of the metabolite in young mice programmed to have mammary tumors as adults, but not elevated in mice that did not get mammary tumors. Bradlow developed a test for 16α-hydroxyestrone in human breast tissue as a potential biomarker for breast cancer.[83] The connection of Bradlow's work on estradiol metabolism and the more recent work on xenobiotics was made when he was contacted by Devra Davis, a public health epidemiologist. Together they recognized that women who lacked most of the known risk factors were afflicted with breast cancer, and they reasoned that if too much estrogen could be dangerous, a woman's prolonged exposure to xenoestrogens could account for some of the cases that heretofore had had no obvious cause. Their collaboration led to an article, published in *Environmental Health Perspectives* in 1993, titled "Medical Hypothesis: Xenoestrogens as Preventable Causes of Breast Cancer."[84] Their argument was structured as follows:

1. Most of the risk factors for breast cancer can be linked to the individual's total lifetime exposure to bioavailable estrogen.
2. Experimental evidence reveals that chemical compounds (xenoestrogens) used in industry and agriculture affect bodily estrogen production and metabolism.
3. Epidemiological evidence reveals that breast fat and serum lipids of women with breast cancer contain significantly elevated levels of some xenoestrogens.
4. Exposures to xenoestrogenic substances may account for increases in breast cancer.

The authors posited a testable biochemical model based on estradiol metabolism, largely developed by Bradlow and his colleagues, to support their conjecture. To understand their model we first need to describe briefly estradiol's role in cell growth. Estradiol binds to an intracellular protein known as the estrogen receptor. Complexes of the hormone and the receptor can bind to DNA in the nucleus and activate genes that direct cell division. The hormone-receptor complexes speed the rate of DNA replication, increasing the likelihood that a mutation will arise and go unrepaired.

The authors described two mutually exclusive biochemical pathways for estradiol metabolism. In one reaction (pathway I) enzymes acting on estradiol produce a substance called 2-hydroxyestrone, and in the other reaction (pathway II) the enzymes produce a substance called 16α-hydroxyestrone. The products differ only in the placement of a hydroxyl (OH) group. Furthermore, these substances cannot be produced at the same time: if more 2-hydroxyestrone is made, less 16α-hydroxyestrone is made, and vice versa. The product of pathway I is minimally estrogenic and nongenotoxic, whereas the product of pathway II is a potent estrogen that is genotoxic. Davis et al. claimed that "substances that elevate pathway II or inhibit pathway I increase risk, whereas those that inhibit pathway II or elevate pathway I decrease risk."[85] Xenoestrogens satisfy the former condition whereas certain diets foster the latter.

This estrogen metabolism hypothesis is consistent with the empirical knowledge that total lifetime exposure to unbound estradiol is one of the known risk factors for breast cancer. It is also consistent with the lower rates of breast cancer among Asian women—even those with high levels of chlorinated organic compounds in their blood—who eat diets rich in soy products, cabbage, broccoli, and other vegetables that increase the levels of 2-hydroxyestrone, thereby lowering 16α-hydroxyestrone levels. According to the Bradlow-Davis mechanistic hypothesis, the 16α metabolite stimulates cell proliferation and allows the new cellular community to grow without attachment to a tissue surface.

Davis and Bradlow jointly authored an update of the *Environmental Health Perspectives* paper in *Scientific American* titled "Can Environmental Estrogens Cause Breast Cancer?"[86] Responding to the widely held criticism that xenoestrogens are not potent enough and are not absorbed in sufficient quantities to cause human health effects, they argued that some of these substances can accumulate and persist for decades in the fat of animals. Davis was also the lead author of a 1997 essay titled "Environmental Influences on Breast Cancer," which discussed the idea of "good" and "bad" estrogens in conection with recent epidemiological data on breast cancer among Asian and African Americans.[87]

The secondary science literature and the popular press began taking note of the new breast cancer hypothesis even prior to the publicaton of the 1993 *Environmental Health Perspectives* article. *Science News* cast the hypothesis as an alternative to the direct mutation theory of breast cancer by emphasizing "more circuitous mechanisms" to explain its rise.[88] Leon Bradlow's 20-year-old research on the metabolic pathways of estrogen and the forthcoming *Environmental Health Perspectives* article were cited as definitive sources for the report. Around the same time, *Ms.* magazine carried a story titled "The Environmental Link to Breast Cancer." Citing Davis, *Ms.* outlined the general scope of the link between endocrine disrupters and cancer: "Certain organochlorines, among them dioxin, DDT, and chlordane, mimic estrogen in the body or interfere with the systems that regulate the production of estrogen and other sex hormones."[89]

A possible connection between breast cancer and xenoestrogens was raised cautiously in *Our Stolen Future.* After quoting Bradlow—"Our data show that a wide variety of pesticides and related compounds clearly have effects on estrogen metabolism that would act in the direction of increasing breast cancer and endometrial cancer risks"—the authors concluded: "Because of our poor understanding of what causes breast cancer and significant uncertainties about exposure, it may take some time to satisfactorily test the hypothesis and discover whether synthetic chemicals are contributing to rising breast cancer rates."[90]

Bradlow continued to pursue opportunities to test his estrogen metabolism thesis. By 1998 several additional studies (human and cell culture) had added support to the postulated relationship between estrogen metabolism and breast cancer risk.[91] One test design still in progress is a case-controlled study (women with breast cancer are compared with women of similar ages without the disease) involving Long Island women. A study of estradiol metabolites in this or other populations of women could help resolve the issue of good versus bad estrogen. Another question yet to be answered is whether women's exposure to ambient levels of xenoestrogens can affect the balance of good and bad estrogen metabolites and thus increase the risk of breast cancer.

Breast cancer has become such a powerful political symbol of unity among women that any association between the disease and chemicals makes chemical manufacturers very uneasy. There is a strong receptivity within the industrial sector to any evidence refuting Bradlow's thesis, and financial backing flows accordingly. Such evidence was purported to have been put forth in a study by Safe and McDougal, who used a series of in vitro tests on estrogen-sensitive MCF-7 cells to argue that production of 2-hydroxyestrone (the good estrogen) can be induced by nonestrogenic and antiestrogenic chemicals and that, as a consequence, the ratio of 2- to 16α-hydroxyestrone is not a predictive assay for putative estrogenic pesticides or mammary carcinogens. The authors also dispute the association between organochlorine pesticides and breast cancer, arguing that they constitute only a small fraction of the daily intake of estrogenic compounds.[92]

Bradlow considered the Safe and McDougal paper fundamentally flawed because it reported data from an assay (developed by Bradlow) that were inconsistent.[93] Safe's role in disputing different components of the hypothesis has also raised eyebrows among some of his colleagues, who consider his industrial funding sources a matter of dishonor in these sensitive areas of science. However, Stephen Safe—who has acknowledged receiving funding from the industry-supported Chemical Industry Institute of Toxicology and the Chemical Manufacturers Association for some of his work—has indicated that he relied on only federal support for the studies reported in the paper that disputes the theory of good and bad estrogens.

Safe was given a platform to advance his views on organochlorines and breast cancer in the editorial pages of one of the world's leading medical journals—the *New England Journal of Medicine.* He was offered the chance to voice an editorial opinion in response to a study published in the same issue that concluded that there was no association between two organochlorines in plasma levels of women and the risk of breast cancer.[94] The journal's editors had an opportunity to yield the editorial page to someone who could point out the limitations of the study or emphasize the difficulty of finding such links. Instead they chose Safe, who cast the study as the definitive case against the industrial link to breast cancer. Safe wrote that this study, along with other studies, "should reassure the public that weakly estrogenic organochlorine compounds such as PCBs, DDT and DDE [1,1-dichloro-2,2-bis ethylene] are not a cause of breast cancer."[95] The *Journal* was subsequently criticized for violating its own conflict-of-interest guidelines in allowing Safe to write its editorial. The response of Jerome Kassirer, the journal's editor-in-chief, to the criticism was described in one press account: "While he didn't know about Safe's CMA [Chemical Manufacturers Association] funding when the editorial ran, that wouldn't have stopped him from publishing it. That's because Safe's funding had stopped several months earlier, he also was getting money from neutral sources like the National Institutes of Health and the U.S. Environmental Protection Agency, and, most important, because CMA funds made up just 20 percent of Safe's research budget."[96]

What should we make of all the scientific sparring over cancer research, such as whether we are asking the right questions, spending money wisely, acting ethically, or getting useful results? For nearly 30 years research scientists have been recruited for combat in the war against cancer, and a significant share of the public's health care research funding arsenal has been directed at that effort. Despite the billions in social investment, some believe we are still no closer to a cure for most terminal cancers, and the incidence rates of many cancers have not diminished. When chemicals are implicated as possible causal agents, the stakes are high for chemical manufacturers. When biochemical or genetic mechanisms of causality are advanced, the stakes are high for the research community. Research dollars and scientific prestige flow abundantly to promising approaches to understanding cancer causation and therapeutic treatment. Members of the research community stake claims to certain cancer paradigms and then engage in a virtual medieval joust to protect the honor of their claims, in combative interactions frequently obscured from public view. Cancer research is, after all, an arena of prestige and privilege among those commissioned to carry out the war.

Among the least privileged and funded combat scientists in the cancer wars are those who investigate environmental causes. For example, the Silent Spring Institute of Newton, Massachusetts, is pursuing a link between chemicals and breast cancer in communities around Cape Cod. Chapter 2 discusses their unique approach in identifying drinking water sources containing estrogenic compounds and in comparing sites with high degrees of exposure with breast cancer rates.

Thus far, the hot pursuit of an industrial chemical link to breast cancer has not yielded definitive results. Yet this outcome has not persuaded the true believers that the accumulation of organochlorines in breast tissue and blood serum is without adverse health effects, and without a satisfactory explanation for breast cancer rates, these findings are all they have. Some analysts, such as Sandra Steingraber, in her book *Living Downstream,* argue that the scientific methods used to detect causal relationships between environmental agents and breast cancer are too reductionist to address the complexity of the chemical interactions. Simply to correlate organochlorine levels in mature women with the risk of breast cancer, according to Steingraber, neglects their history. Those who were breast-fed may be at higher risk; those who breast-fed their infants may have driven pollutants from their bodies into those of their children. According to Steingraber, "Because cancer is a multicausal disease that unfolds over a period of decades, exposures during young adulthood, adolescence, childhood—even prior to birth—are relevant to our present cancer risks."[97]

The wildlife literature did not reveal any important clues about the relationship between xenoestrogens and breast cancer. However, wildlife research did provide insights into the effects of chemicals on brain function and behavior. Moreover, it turned scientific attention to in utero chemical exposures and their impacts through the first few years of life. Theo Colborn provided the stimulus for investigating these themes in a workshop that brought together new communities of scientists to address the effects of endocrine disrupters on neurological and behavioral development.

Behavioral and Neurophysiological Effects

The relationship between chemicals and the brain was on Colborn's agenda from the early stages in the development of her views on endocrine disrupters. In the first Wingspread Work Session she invited several speakers to discuss disturbances in neuroendocrine development, including sexual orientation and behavior. The consideration of endocrine disrupter effects on the brain of the developing fetus became a more focused part of Colborn's inquiry in 1993, when she and her collaborators were working on a framework for *Our Stolen Future*. In the summer of 1994 she and Dianne Dumanoski traveled to Oswego, New York, to meet with Helen Daly of the Center for Neurobehavioral Effects of Environmental Toxins at the State University of New York at Oswego, who had investigated the effects of feeding contaminated salmon to rats.[98] Dumanoski describes the influences that shaped her and Colborn's views on the cognitive and behavioral effects of endocrine disrupters:

> There were two catalysts: a paper by Peter Hauser on ADHD [attention-deficit/hyperactivity disorder], showing that those with an inherited form of this disorder have defective thyroid receptors which make them resistant to the hormone. . . . Then I got a paper by Susan Porterfield, a researcher who looks at the role of thyroid and brain development. This was all going on while we were developing the "Altered Destinies" chapter [in *Our Stolen Future*]. As we were doing this work, I became persuaded that thyroid disruption and brain and behavior ultimately might prove to be far more important than reproductive effects and I made the case that we should elevate more attention to this than we had initially planned.[99]

In 1995, Colborn organized a workshop in Erice, Sicily, titled "Environmental Endocrine-Disrupting Chemicals: Neural, Endocrine, and Behavioral Effects" and sponsored by the Italian government and several foundations. This meeting focused heavily on neuroendocrinology and thyroid disorders. And like the Wingspread work sessions that had preceded it, the meeting resulted in a consensus statement, known as the Erice statement, which contributed to a deepening of the hypothesis by bringing together scientists who study the human cognitive and behavioral effects of chemicals with wildlife toxicologists.

The inclusion of cognition and behavior within the scope of the hypothesis came about when it was recognized that during the first few weeks of gestation some environmental chemicals can interfere with the hormone thyroxin (T-4), which is produced in the thyroid gland and is necessary for normal brain development. Low levels of thyroxin must enter the fetal brain for an individual to develop normally. If thyroxin's action is blocked, it is believed that various gradations of lower cognitive function—such as loss of intelligence, mental retardation, and abnormal behavioral outcomes—will result. Many of the chemicals implicated as endocrine disrupters also interfere with thyroxin production, with thyroxin transport, with the enzyme that converts one form of thyroxin (T-4) to another (T-3), or with the binding of chemicals to thyroid receptor sites. PCBs, certain pesticides, and dioxin are all capable of interfering with thyroid hormones.

The relationship between endocrine disrupters and behavior was strengthened when Peter Hauser of the National Institutes of Health published an article in the *New England Journal of Medicine* reporting that among children with ADHD a number had a genetic deficiency called thyroid hormone resistance.[100] As a result of this deficiency they were not responsive to thyroid hormone. According to Colborn and others at the Erice meeting, there are good reasons to believe that interference with thyroid hormone's role in brain development during the early stages of pregnancy might contribute to ADHD.

Adverse changes in intelligence and behavior due to prenatal exposure to minute quantities of endocrine disrupters, although difficult to prove, have serious implications. A central tenet of the hypothesis is that exposure of wildlife and humans to chemical contaminants during prenatal life can lead to "profound and irreversible abnormalities in brain development at exposure levels that do not produce permanent effects in adults."[101] Cognitive and behavioral effects are more difficult to document than physical abnormalities or clinical disease, but they could nonetheless have a significant impact on public policy. For example, in the last decade researchers have associated early exposure to lead, at levels until very recently considered safe, with lower IQs in two-year-old children.[102] These findings, along with other studies on childhood lead exposure, contributed to the banning of leaded gasoline in the United States and to the passage of new requirements for the removal of lead paint from homes. In addition to lead, other chemicals that have been identified as being thyroid disrupters are PCBs, polybrominated biphenyls, dioxins, and hexachlorobenzene, although it is not known by what mechanism they affect cognitive function.

In both animal and human studies researchers have examined the effect on cognitive ability and behavioral functioning of early exposure to organochlorine compounds that act as endocrine disrupters. Some of these chemicals are persistent and lipophilic (they have an affinity for fats and oils). During prenatal and neonatal development all animals are exposed to synthetic chemicals because these substances are stored in their mothers' fat and serum. Developing fetuses are exposed either in the egg (birds, reptiles, and fish) or through the umbilical cord and their mothers' milk while breast-feeding (mammals). The global extent of exposure of developing fetuses and neonates to contaminants was first recognized in the 1960s. The levels of exposure are normally relatively low, so that pregnant women and their offspring show no acute signs of toxicity. But researchers have wondered for years about the possible impacts of these early exposures on cognitive and behavioral functioning throughout a lifetime. Are the intellectual functioning and normal behavioral adjustment of our children being compromised because of chemical contamination at the dawning of life? Could the perceived increase in ADHD in the population—which is often attributed to such cultural changes as poor diet, excessive television watching, and the increasingly frenetic nature of modern life—be in part the result of early exposure to harmful and ever-present chemicals that disrupt normal brain development?

Two mass poisoning events in Asia provide evidence supporting the link between high levels of exposure to certain organochlorine compounds and developmental disorders. In 1968 in Japan and in 1978–79 in Taiwan, consumers were exposed to rice oil contaminated with PCBs and other organochlorine compounds. The people consuming the tainted oil became sick with what became known as "oil disease" (*Yusho* in Japanese and *Yu-Cheng* in Chinese). Because the chemicals persist in humans, children born years after the incidents continued to be exposed in utero and during breast-feeding. In Taiwan more than a hundred children born to the Yu-Cheng women up to six years after the incident have been followed.

Many of these children show physical signs of the poisonings, including small size and weight, hyperpigmentation, nail deformities, and lung problems.[103] In addition, the Yu-Cheng children consistently score lower on cognitive tests and more poorly on behavioral assessments than controls or a group of older siblings that had not been exposed. Children born up to six years after exposure were just as affected as those born within a year of exposure, and the effect continues at least six to seven years after birth.[104]

The Yu-Cheng children were exposed to PCBs at levels high enough to cause a variety of acute physical symptoms as well as cognitive and behavioral effects. But a central premise of the endocrine hypothesis is that the general population is at risk from chronic, perhaps unavoidable, exposures that are not associated with acute effects. There have been two large studies in the United States to assess subacute exposures to PCB complexes: one in Michigan and another in North Carolina.

Well before the release of the Wingspread I statement, researchers studying the Great Lakes region expressed an interest in the effects of xenobiotics on wildlife and humans because of the high levels of organochlorine and heavy metal contamination in the aquatic food chain. Joseph and Sandra Jacobson and their colleagues at Wayne State University discovered adverse cognitive effects of PCBs in children who appeared healthy, responsive, and normal at birth. The mothers of the affected children did not show any acute clinical symptoms of PCB poisoning. Of the mothers in the study, 75 percent had consumed at least 26 pounds of fish from Lake Michigan, which contained PCBs and other xenobiotics, in the six years prior to the birth of the children in the study. The other 25 percent of the mothers reported they had not consumed these fish.[105]

The Wayne State University investigators identified adverse cognitive effects from prenatal PCB exposure by the time infants reached seven months of age. There was a significant association between a baby's in utero exposure to PCBs, measured by umbilical cord serum levels, and lower scores on a visual recognition intelligence test.[106] The test measures the relative response of the baby to a novel object compared with a familiar one. Higher scores, which may be a predictor of later intelligence, result when babies visually prefer the novel object. Interestingly, there was not a significant association between exposure to PCBs neonatally from breast-feeding (the concentration of PCBs in breast milk is much higher than that in umbilical cord serum) and visual recognition intelligence test scores. This finding suggested that cognitive deficits from PCB contamination were the result of fetal exposure rather than postnatal exposure from breast-feeding.

When children in the study were tested at four years of age, certain cognitive deficits associated with prenatal PCB exposure were also found. Children with higher prenatal PCB exposure had significant deficits in short-term memory function and cognitive processing speed compared with children having lower prenatal exposures. Levels of infants' exposure to PCBs during nursing and their current body burdens of PCBs (which are highly correlated to each other but not to PCB levels in umbilical cord serum) were found to be unrelated to cognitive functioning.

The North Carolina study, which was conducted on a sample of the general population, found psychomotor developmental deficits in children with higher levels of prenatal exposure to PCBs at six and twelve months.[107] As in the Michigan study, exposure of children to PCBs from breast-feeding had no detectable developmental effect.

Several factors help explain why fetuses might be more sensitive to exposure to xenobiotics than newborns or young children.[108] Detoxification mechanisms do not develop until after birth. Young children have fat deposits that are lacking in fetuses, and these deposits can absorb xenobiotics. The brains of fetuses may also be more vulnerable to xenobiotics because the blood-brain barrier does not form until later in development. Finally, dividing and migrating cells in fetuses may be especially sensitive to their chemical environment, much more so than cells in neonates. Although no mechanism has been proposed to account for the cognitive and developmental effects associated with exposure to PCBs, all of these factors suggest increased vulnerability of fetuses to low-level xenobiotic exposure, which is consistent with other findings associated with the environmental endocrine hypothesis.

Scientists began to investigate whether exposure to low levels of environmental contaminants could have effects on behavior, activity levels, or cognitive function in humans. A series of laboratory studies conducted by Helen Daly showed that rats fed fish containing high levels of xenobiotics had altered behavior compared with control rats fed no fish or fed uncontaminated fish.[109] Rats fed fish from Lake Ontario behaved similarly to control rats in tests using positive reinforcements. However, when a negative experience was introduced as well as the reward, the test rats became easily frustrated and tended to give up more easily compared with the control rats. Daly describes this as "hyper-reactivity to unpleasant events." Most interestingly, the offspring of the test rats, who themselves had never been fed the contaminated fish, exhibited the same behavioral changes as their mothers. Daly contended that this experimental evidence could provide a "missing link" to behavior noted in human studies.

Evidence has been mounting that environmental chemicals, particularly endocrine disrupters, can lower cognitive functions and affect behavior. The effects that have been documented to date appear to apply significantly but not exclusively to those individuals exposed prenatally, a finding that fits with the general emphasis placed on in utero effects by proponents of the hypothesis. However, studies that have found neurobehavioral effects have not discovered their underlying causal mechanisms. It remains unresolved whether interference with normal endocrine functioning essential to brain and nervous system development, or some other mediating influence, explains the observed cognitive and behavioral abnormalities.

Many of the scientists who attended the Erice meeting were not familiar with the links that Colborn had postulated between environmental chemicals and hormone modulation, and therefore a segment of the meeting was devoted to bringing the new scientific cohort up to speed. Since the neurosciences is a highly respected area of biomedical research, the consensus statement from this meeting could do nothing but elevate the stature of the general endocrine hypothesis. The consensus statement from Erice was signed by 18 participants and published in the journal *Toxicology and Industrial Health.* Its first paragraph stated:

> We are certain of the following: Endocrine-disrupting chemicals can undermine neurological and behavioral development and subsequent potential of individuals exposed in the womb or, in fish, amphibians, reptiles, and birds, the egg. This loss of potential in humans and wildlife is expressed as behavioral and physical abnormalities. It may be expressed as reduced intellectual capacity and social adaptability, as impaired responsiveness to environmental demands, or in a variety of other functional guises. Widespread loss of this nature can change the character of human societies or destabilize wildlife populations.[110]

Statements of support for the signatory scientists were also disseminated through the Environmental Information Center's home page on the World Wide Web. In one of these letters, Bernard Weiss, professor of environmental medicine at the University of Rochester, wrote:

> In Erice, a group of scientists from diverse disciplines and backgrounds came together to confront another kind of chemical hazard. It

comes in the form of a biological domain that I and many of my colleagues in toxicology previously had neglected. No longer. We now know that some agents, including many we had investigated in other contexts, can act as hormone mimics or antagonists and exert profound effects on brain development. Their consequences for human welfare won't be fully grasped without an immense effort in the laboratory, in the clinic, and in the field. We live in interesting times.[111]

While the Erice consensus position was circulating within and beyond the scientific community, giving more structure to the general hypothesis of endocrine effects, a subsequent study published by the Jacobson team provided stronger epidemiological evidence that prenatal exposures can adversely affect the cognitive function of children beyond infancy. On September 12, 1996, the *New England Journal of Medicine* published their findings on the effects of children exposed to PCBs in utero. The investigators tested 212 children selected as newborns in 1980 and 1981 and followed them at age 11. Of the children, 187 had been born to women who had eaten Lake Michigan fish contaminated with PCBs. The study found lower than average IQ scores, poor reading skills, memory problems, and attention deficits in the children prenatally exposed to PCBs in concentrations only slightly elevated above those found in the general population.[112] In its coverage of the Jacobson study, the *New York Times*—which had by this point covered several stories on endocrine disrupters—curiously neglected to connect the results to the general endocrine hypothesis, instead treating the results as an isolated case of a chemical health effect.[113] This was consistent with the skepticism expressed in past coverage by the *Times* of *Our Stolen Future* and the overall theory of endocrine disrupters.

Early Evidentiary Support

Between 1987 and 1991, with the help of a committed group of scientific collaborators, Theo Colborn assembled the initial body of evidence in support of her general thesis on endocrine-disrupting chemicals. The next period in the scientific development of the endocrine hypothesis was marked by the first Wingspread Work Session (1991) and the publicaton of *Our Stolen Future* (1996). During that time Colborn cultivated and nurtured a science constituency that helped her develop a deeper scientific understanding of endocrine disrupters and broaden the evidentiary support for her findings. In those years she was instrumental in organizing six workshops (three in 1995 alone), each resulting in a consensus statement on the role that endocrine disrupters are known or suspected to play in a variety of human and wildlife abnormalities.

How strong was the evidence for the role of endocrine disrupters during this period, when the issues were just starting to interest policy-makers and the popular book had not yet been released? One way to answer this question is to examine the evidence as seen through scientific review articles. Reviews in science are usually solicited by journal editors, government agencies, or other leaders in a field in the interest of reporting the current state of scientific understanding on a given issue. Review articles provide a window into the development of a scientific field through the eyes of the scientific gatekeepers. No two authors are likely to select the same body of work for review.

Between 1992 and 1995 only a handful of review articles were written on endocrine disrupters: a literature search revealed only seven, covering both wildlife and human health effects. Of the four reviews that discussed wildlife data, two were published in the proceedings of the first Wingspread Work Session,[114] one appeared in the journal *Environmental Health Perspectives,*[115] and a fourth was a report by the Danish Environmental Protection Agency.[116] Since the science of endocrine disrupters, particularly as it applies to wildlife, was neither well known nor strongly contested during this period, these reviews represent a body of evidence put forth in support of the hypothesis by people like Colborn who had already been persuaded of its cogency. An analysis of the four wildlife reviews reveals the following. Scientists had assembled evidence of endocrine effects of specific industrial chemicals, pesticides, and sewage discharge across eight taxa—fish, mammals, birds, reptiles, echinoderms (e.g., sea urchins), worms, molluscs, and zooplankton—examined in 70 studies. The largest number of citations came from studies of birds (20) and fish (16), followed by molluscs (13) and mammals (11). The dates of the published studies reviewed began with 1970 and continued through 1995.

The agents most frequently identified as the cause of a pathology were various pesticides (including DDT and its metabolites) and PCBs. The extensive field work on molluscs over a 10-year period connected their exposure to tributyltin (a chemical used in shipbuilding) to imposex—the abnormal development of male characteristics in female marine organisms. The most prevalent site locations for field studies were the Great Lakes and the Channel Islands off California.

Six of the seven review articles discussed the evidence of human health effects of endocrine disrupters. Three appeared in the journal

Environmental Health Perspectives;[117] one came from an unabridged unpublished version of a paper published in *Scientific American;*[118] one was published in the proceedings of Wingspread I;[119] and finally, an extensive literature review of male reproductive health was contained in a report from the Danish Environmental Protection Agency.[120]

Approximately one-third of the studies involved humans, whereas the majority involved animal models or chemical assays. The four major pathologies cited were testicular dysfunction and semen impairment (54 studies in males); abnormal genital tract development, vaginal cancer, and infertility in women (21 studies in females); breast cancer (16 studies in females); and cognitive and behavioral abnormalities (8 studies in both males and females). Studies were cited from 1969 through 1995. Where suspected causal agents were identified, DES, PCBs, DDT, and other organochlorine pesticides were the ones most frequently cited. On such subjects as semen quality and breast cancer, the human studies were not consistent. Some of the strongest associations were found in occupational studies and in studies performed on neonates.

The seven review articles published during the period 1992–95 cited 151 studies, most of which are either consistent with or show evidence of a chemically induced effect on the endocrine system of an animal or human. The studies are spread across a broad spectrum of animal and human effects. In some ways this strengthens the impact of the general hypothesis, but in other respects the evidence is diffuse and unfocused. The next period in the development of the endocrine hypothesis would be that of focusing on specific outcomes and deepening investigations into the mechanisms through which endocrine disrupters operate.

In the research arena of biomedical science, success is measured in part by the degree to which one's theory of causation provides the framework for new areas of investigation. Many investigators make small and meaningful contributions, but few can claim that their research has opened up new paths of inquiry or contributed a new lens through which to interpret published research.

The evolution and development of the environmental endocrine hypothesis are noteworthy for many reasons, ranging from the multiple pathways of discovery to the responses of various scientific disciplines after the hypothesis was formulated. Certainly it was unusual that a newly minted Ph.D., without a university affiliation and employed by a public interest environmental organization, advanced the most generalized form of the hypothesis. Yet perhaps we should not be too surprised at this outcome. After all, Einstein wrote the paper outlining his theory of relativity while employed in the Swiss patent office. Theo Colborn's creative contribution was in the discovery of patterns. She kept seeking the forest while everyone else was still looking at the trees.

But there are always two phases to any significant discovery. There are the personal intuitive and intellectual insights—the "Eureka!" phenomenon. And there is also the social dimension, through which a discovery is used to convert one's peers to an acceptance of a new theory. Science and its social context are not isolated in this process. There is a reciprocity. Once the issue of endocrine disrupters had been addressed by Congress, scientists began to take notice, particularly when research dollars were made available and certainly when legislation was enacted. In addition, the public visibility of a possible explanation for preventable human health problems can raise the general level of literacy among previously quiescent segments of the scientific community.

The next chapter explores the emergence of a public side to the new theory of chemically induced disease during a period of more accelerated and focused scientific inquiry.

CHAPTER

2

The Emergence of a Public Hypothesis

Scientific discoveries that reveal public health risks often have protracted gestation periods of internal debate before they are recognized and acted upon in the policy arena. Many suspected occupational hazards of chemical exposure (such as exposure to asbestos and vinyl chloride) lay dormant for decades, unrecognized by public health agencies, because medical records were held by company physicians. Similar examples can be found for environmental agents. The chemical ethylene dibromide (EDB) has been used commercially since the 1920s, first as a gasoline additive and then as a soil fumigant for nematode control and as an insecticide against fruit flies and a fungicide for stored grain. It was registered as a pesticide in 1948. In 1974 EDB was identified as a cancer-causing agent. Ten years later, it was not the scientists and not the federal government, but state officials who, by rejecting federal tolerance levels and issuing public health advisories, were responsible for bringing the risks of EDB residues to public attention. These advisories resulted in one of the most intense and dramatic environmental media events in modern history, as cake mixes and children's cereals were pulled from supermarket shelves because they contained high levels of EDB residues. For the EDB case, the media played a central role in turning the contested science of a risk estimate into a full-fledged policy debate.[1]

What was the path followed by the scientific hypothesis of endocrine disrupters to become a public hypothesis? By *public hypothesis* I mean that stage in the development of a scientific hypothesis during which segments of the public feel they have a stake in the outcome of the scientific debates and therefore make increasing demands in order to establish a clearer understanding of the conflicting views.

There is nothing logical or predictable about the way in which scientific dialogues over potential hazards become incorporated into social discourse. The events that turn a scientific issue into a sustained public debate are often erratic and highly particularized. Suspected hazards may lie dormant for years before, suddenly and unexpectedly, they are brought under public scrutiny. My focus in this chapter is on those telltale events that stand out against the continuous background activity of scientist-activists, public officials, and journalists. These include major media events, scientific discoveries and breakthroughs, and notable policy decisions. I provide a picture of activities related to endocrine disrupters that broke through the threshold of inattention and eventually elevated the hypothesis into the policy arena both nationally and internationally.

From 1987 through the mid-1990s the environmental endocrine hypothesis had been circulating among various scientific affinity groups, picking up sympathetic followers (a number of whom had attended the Wingspread workshops), who expressed at best cautious support for the hypothesis in their publications and public statements. During that period *Environmental Health Perspectives* was the only journal that published major review articles covering the full breadth of the hypothesis.[2] Congressional hearings in 1991 and 1993 brought some limited awareness of the potential hazards of endocrine disrupters to the policy and environmental advocacy sectors, but those hearings had little direct effect on the general public. The media paid scant attention to the issue through 1993. It was not until 1994 that significant news coverage began to appear on the postulated connection between synthetic chemicals and endocrine-mediated disorders.

From studies of public perception and societal selection of risks a number of factors have been identified that help explain why certain risk hypotheses become amplified in the public mind and elevated to the social agenda.[3] Among them are a catastrophe or perceived catastrophe, a consensus reached among leading scientists on a human health hazard, an influential media event that shapes public opinion, a seminal discovery, or the publication of an influential book. The rising importance of the environmental endocrine hypothesis in the public arena was influenced by several mutually reinforcing events that took place over a period of several years. No single event can fully explain how the issue was transformed from a scientific curiosity into the subject of a mature public policy debate. The factors contributing to the high visibility of endocrine disrupters during the mid-1990s included federal agency interest (particularly on the part of the Environmental Protection Agency [EPA], fostered by congressional initiatives in the form of hearings and committee reports); a British Broadcasting Corporation (BBC) documentary on falling sperm counts; concern on the part of breast cancer activists about environmental estrogens; and the publication of a mass-market book. The 1996 release of *Our Stolen Future,* followed by extensive media coverage of its message, established endocrine disrupters as a new environmental threat. A National Academy of Sciences (NAS) study panel, contracted by two federal agencies, gave legitimacy at least to the questions raised by the hypothesis.

Finally, the issues associated with the hypothesis were beginning to find a place on the research agendas of funding agencies. Research on reproductive toxicity, heretofore hardly a thriving field, could now be placed on a preventative health agenda, thus achieving a broader reach. Additional research funds to study different components of the environmental endocrine hypothesis soon became available. Scientists in a variety of subfields of molecular and cellular biology, toxicology, and environmental sciences, taking notice of the new funding opportunities, began to reorient their model systems to compete for a share of the newly available grant money. Once it enters America's network of biomedical and environmental funding streams and is incorporated within program requests for proposals, a scientific hypothesis gains new constituencies. Investigators take seriously the extant literature, since reference to it must be part of their grant proposals.

As noted previously, these funding and legislative initiatives had as their starting point congressional hearings in the early 1990s.

Congressional Excitations

Two congressional committee hearings set the stage for federal involvement in the regulation of xenoestrogens. A Senate hearing in 1991 titled *Government Regulation of Reproductive Hazards* featured testimony from Theo Colborn, who, during that period, was a senior fellow with the World Wildlife Fund, an international environmental organization.[4] A second subcommittee hearing in the House two years later titled *Health Effects of Estro-*

genic Pesticides brought testimony from scientists who had participated in the first Wingspread Work Session.[5]

The Senate hearing was run by John Glenn of Ohio, who chaired the Committee on Governmental Affairs. Also on the committee were Senators Carl Levin of Michigan and Herbert Kohl of Wisconsin, representing two Great Lakes states affected by years of unregulated dumping of industrial pollutants. In his introductory remarks Glenn referred to the human health effects of polychlorinated biphenyls (PCBs) in the Great Lakes:

> During the 1980s, a multi-year study found that the amount of Great Lakes fish consumed by pregnant women directly corresponded to the amount of PCBs found in their tissues, breast milk, fetuses, and in their children. Children with greater amounts of PCBs at birth were found to have lower birth weights, smaller head sizes, and slower reaction times. At 4 years of age, these children continued to display developmental delays and reduced motor functions. Similar effects have also been observed in the wildlife dependent upon foods derived from the Great Lakes basin ecosystem.[6]

Noteworthy in his remarks was the chairman's emphasis on children's health. Perceived risks to young children—as evidenced by the Tylenol, Alar, and automobile airbag cases in recent years—generally foster strong and sympathetic public responses. However, endocrine disrupters represented only one part of the subcommittee's purview. Testimony focused on many sources of reproductive hazards, including drugs, consumer products, occupational exposures, and industrial pollutants. The Senate hearings were timed to coincide with a soon-to-be-released study by the General Accounting Office titled *Reproductive and Developmental Toxicants: Regulatory Actions Provide Uncertain Protection.* When the study was submitted to the Senate in October 1992, it had identified 30 chemicals most widely recognized as reproductive and developmental toxicants, and it emphasized that the current laws provided uncertain protection against those substances. According to the report, "When chemicals are regulated, that regulation is seldom based on reproductive and developmental toxicity."[7] However, it included no references to endocrine disrupters or to the scientific papers of the Wingspread participants. The General Accounting Office study marks an important transition point. Thereafter, federal agencies reporting on reproductive and developmental anomalies could not avoid framing these issues, at least in part, in terms of endocrine disrupters.

At the Senate hearing, Theo Colborn was the sole witness representing both the public interest community and scientists working on the health effects of endocrine disrupters. Colborn referred back to her 1987 work, undertaken for the Conservation Foundation, which had led to her book *Great Lakes, Great Legacy?* Based on this work she had been asked by the Institute of Medicine, a unit of the NAS, to write a paper on the health effects of Great Lakes contamination. That paper, titled "Nontraditional Evaluation of Risk from Fish Contaminants," incorporated some of the seminal ideas that would ultimately evolve into her theory of endocrine disrupters. Colborn wrote in her abstract: "This paper argues for shifting the emphasis from cancer, chronic and acute effects to more subtle health decrements when considering the hazards of eating contaminated fish. This position is based on evidence from wildlife, laboratory, and human research."[8] She argued that it is unwise to focus only on cancer and on chronic and acute toxicity when determining the risk of eating contaminated fish. Other effects to be considered include alterations in development, growth retardation, and subtle changes in behavior.

The hearing afforded Colborn an opportunity to report to Congress the conclusions and recommendations in the Wingspread consensus statement, which had grown out of the workshop held three months earlier. Colborn displayed the political acumen of a seasoned Washington activist. She turned the scientific consensus position into a policy argument—one that made an impact on the committee. She noted that it had been concluded at Wingspread "that some of the developmental impairments reported in humans today are seen in offspring that have reached adulthood of parents exposed to environmental chemicals that are hormonally active, and unless the environmental load of these chemicals is abated and controlled, large-scale dysfunction at the population level is possible."[9]

Society, she declared, had been fixated on cancer mortality, gross birth defects, and acute toxicity to the exclusion of fetal exposures. Colborn underscored the point that humans and wildlife are exposed to chemicals below levels at which we see cancer or the other effects. Nevertheless, these exposures are capable of affecting the normal development of vital organs.

The reception to her testimony was very positive, and the reaction of subcommittee members was not one of skepticism. Instead, they asked how bad things really were, whether the effects were irreversible, and how Congress should proceed. At the hearing, Colborn put forward a set of policy recommendations that included broadening public awareness of hormonally active environmental chemicals, testing chemicals with hormonal activity over several generations of animals, developing a major research initiative for the study of the effects of these chemicals in humans, and establishing new approaches to reducing human exposure. Within two years after that first Senate hearing, Congress began acting on these recommendations. Within five years it had reached a decision on its initial action plan.

Two sources of supporting scientific evidence were highlighted in Colborn's testimony: Michigan fish eaters and pregnant rats. She cited the studies by Jacobson et al. that showed that pregnant women who had been high consumers of fish gave birth to children with physical and cognitive deficits.[10] These studies provided the clearest indications of the effects on human health of the PCBs that had contaminated the Great Lakes. Colborn also highlighted studies of pregnant rats exposed to one meal of dioxin that resulted in measurable demasculinization and feminization. The stories of disrupted sexual development in wildlife and laboratory animals resulting from fetal exposure to xenobiotics stood in striking contrast to society's single-minded obsession with cancer.

Under questioning, Colborn introduced a powerful dialectical idea: chemicals can exhibit distinct, sometimes opposite, effects during different phases in the life cycle of an organism. "Dioxin in a fetal organism has an estrogen-like effect. In adults, it has an anti-estrogen effect."[11] This is one of the key factors that makes endocrine disrupters more complex in performing a risk assessment than carcinogens or acute toxins. It is understood that chemical carcinogens are capable of inducing genetic mutations in cells. There is no ambiguity about "good" and "bad" carcinogens. However, critics of the environmental endocrine hypothesis would later exploit the ambiguity of good and bad estrogens, or estrogens and antiestrogens, to raise uncertainties about the seriousness of chemical hazards. If a chemical behaves sometimes like an estrogen and sometimes like an antiestrogen, how can we infer a definitive cause-effect relationship in humans? In contrast to that of chemical carcinogens, the study of endocrine disrupters was complicated by the observation that a hormonal impact on animals depended on when in the animal's life cycle exposure had taken place. There were windows of vulnerability for endocrine-disrupting effects.

Henry Waxman of California chaired the House Subcommittee on Health and the Environment in 1993. The context of the subcommittee hearings was a new bill, the Food Quality Protection Act of 1993 (H.R. 1627), one of the goals of which was to speed up the pesticide re-registration process. Many pesticides then in use had not been evaluated properly when they received their initial registration or had been grandfathered into use. Pesticide manufacturers had been required since 1972 to re-register their products, yet re-registration had slowed to a snail's pace. Another goal of the act was to supersede the Delaney clause in the Food, Drug and Cosmetic Act of 1958 in an effort to resolve legal wrangling over regulating the level of pesticide residues in food.* Agribusiness wanted the Delaney clause repealed as it applied to pesticide residues in processed food. Environmentalists wanted to retain it unless they could be assured that its repeal would be balanced by a net gain in safety, with special considerations given to such vulnerable populations as infants.

Whereas the British and Danish reports on the endocrine effects of chemicals focused almost exclusively on male reproduction and xenoestrogens, Waxman's hearings emphasized the impacts on both human and wildlife health of estrogen-imitative compounds in the environment. In his introductory remarks, Waxman cited three important findings from the hearings. First, he pointed out that estrogenic chemicals might be contributing to rising breast cancer rates in the United States, a dramatic worldwide decline in the male sperm counts over the last 50 years, and reproductive failures in wildlife species, including alligators in Florida and bears in Alaska. Second, he noted that estrogen mimics, primarily from pesticide residues, are found in the food supply. Finally, Waxman emphasized that the EPA does not routinely evaluate pesticides for dangerous hormonal effects.

During a joint House-Senate committee hearing on the safety of pesticides and food held on September 21, 1993, and chaired by Waxman, Oklahoma Congressman Mike Synar decried the snail's pace of pesticide testing: "Almost 20,000 pesticide products have been under review since 1972 and only 31 have been re-registered. At this rate it will take us to the year 15,520 A.D. to complete. I believe in flexibility. I believe in good science. What I don't believe in is geological time."[12] The connection of endocrine disrupters to breast cancer in the early 1990s was curious in one respect, since the evidence for a xenoestrogen cause of breast cancer was quite speculative. Nevertheless, the breast cancer connection was a strong focus of the Waxman hearings, and it continued to frame the discussion of endocrine disrupters for a number of years. Of the 17 presenters, 8 addressed breast cancer as a primary area of concern. Among

*The Delaney clause states that any substance shown to be carcinogenic when ingested by humans or animals is prohibited for use in processed food.

those testifying was Devra Lee Davis, adviser to the Department of Health and Human Services, who, with Leon Bradlow of the Cornell Medical School, had postulated a metabolic hypothesis for the relationship between xenoestrogens and the increased risk of breast cancer (see Chapter 1).

Waxman understood the alarm that media reports of his hearings could create in the public. He began the hearings with a sober reminder of the need to avoid consumer panic. The Alar episode—in which media investigations into the safety of the chemical growth agent resulted in overnight boycotts of apples and apple products—was still imprinted in the consciousness of policymakers: "The potential adverse health effects from hormonal pesticides are alarming, but they are not yet proven. The evidence about hormonal effects is not yet conclusive enough to warrant changes in food consumption. In particular, families and other consumers should continue to eat a plentiful diet of fruits and vegetables, which have many positive health benefits. Instead of consumer panic and changes in diet, what we urgently need is more investigation and crucial regulatory improvements."[13]

Wingspread alumni were heavily represented among the scientists who testified; they included Theo Colborn, Earl Gray, Louis Guillette, John McLachlan, and Ana Soto. Yet even among this group of scientists—who had invested much effort and thought in considering the human health impacts of environmental endocrine disrupters—there were differences on such critical issues as the degree of confidence that endocrine disrupters were indeed having an effect on humans. McLachlan ended his testimony cautiously: "It is not possible at this time to confirm a cause and effect relationship between environmental estrogens and human disease or dysfunction."[14] Bench scientists like McLachlan, accustomed to a methodology that manipulates a single variable in a system and measures the outcome against a control, are characteristically cautious in their use of causal language. Rarely, if ever, is this methodology successful in discovering human effects of chemicals.

Colborn, on the other hand, approached the issue of human effects from exposure to xenoestrogens with less reserve than she had shown in her 1991 testimony. While hesitating to use causal language, she drew an analogy with observations in wildlife populations: "Our biosphere is saturated with chemicals that cause functional disorders. Humans are now carrying burdens of both industrial and agricultural chemicals at concentrations at which adverse endocrine, immune and reproductive effects have been reported in affected wildlife and laboratory animals. There is growing evidence that some of these humans have also been affected as a result of their parents' exposure to endocrine disrupting chemicals."[15]

Other scientists expressed doubt as to the certainty of a causal connection. Mary Wolff of the Mount Sinai School of Medicine stated that "much remains to be done to confirm the findings linking organochlorines to breast cancer and to clarify the underlying mechanisms." Louis Guillette, a reproductive endocrinologist at the University of Florida, testified that he had more than correlational evidence that 1,1-dichloro-2,2-bis ethylene (DDE, a metabolite of dichloro-diphenyl trichloroethane, better known as DDT) was responsible for a decline in alligator populations. He spoke of having data that supported "the hypothesis that contamination of a wildlife population with an estrogenic compound has the potential to significantly alter embryonic sexual development and thereby significantly depress subsequent reproductive success." Even though he had been able to reproduce the effects of wildlife exposure to certain chemicals in a laboratory study, he resisted the use of causal language in describing the alligator decline. Earl Gray also used laboratory studies of animals to draw the analogy with humans. Here again, the tentativeness of the causal linkage was emphasized: "It is likely, but not proven, that the active metabolites of vinclozolin [a pesticide] would also bind to the human androgen receptor, and could produce adverse effects on human male sex differentiation." The task, Gray added, was to determine how the doses given to laboratory animals related to doses to which humans were exposed.[16]

The hearing ended with a recognition of three areas of uncertainty. The mechanism of action through which endocrine disrupters establish their effects was still being investigated. Human exposure and its relationship to the dose-responses of animals had yet to be determined. Finally, the additive effects of multiple endocrine disrupters stored in body fat were largely unknown. It was also acknowledged that a generalized regime of bioassays was needed to prescreen chemicals for endocrine effects before they were introduced into industrial or agricultural use and to identify those endocrine disrupters that were already in use in commerce.

Given uncertainties of the kind expressed at this hearing, one might have expected Congress to move rather conservatively—perhaps to fund additional research but delay the passage of legislation until the human effects of endocrine disrupters were better understood. However, social pressure was building in favor of governmental action in some form. Contributing to the milieu was an unusually provocative television documentary produced in the United Kingdom that found its way into influential U.S. policy circles.

A British Documentary

The first major media event highlighting the broad reach of the environmental endocrine hypothesis was a BBC documentary released in 1994 titled *Assault on the Male.* As suggested by its title, the film outlined the various scientific hypotheses that linked chemicals to reproductive and developmental abnormalities in males.

The broadcast was shown in the United States on the cable Discovery Channel but failed to reach the broader viewership on public or network television. The award-winning *Nova* series, produced under the auspices of Boston public television station WGBH, had a co-production agreement with the BBC for several films a year. *Nova*'s producers typically would select a film under the agreement before it was finished. In the case of *Assault on the Male, Nova*'s science editors received a finished version of the film and, following their usual practice, sent it out for review. WGBH places a high premium on accuracy and balance in its *Nova* broadcasts. When their reviewers' comments indicated that the film had a distinct point of view, the editors realized it would take considerable work to achieve the balance they sought. With other films to choose from to satisfy the co-production agreement, *Nova* rejected *Assault on the Male,* whereupon the BBC sold the air rights for the documentary to the Discovery Channel.

The Discovery Channel aired the film in September 1994 on Sunday at 7 P.M. and Monday at 2 A.M. during the Labor Day weekend. The Nielsen rating of 0.6 for the 7 P.M. showing was below average for the Discovery Channel and represented about one-fiftieth the rating of a typical prime-time program. The broadcast of *Assault on the Male* is estimated based on this Nielsen rating to have reached 370,000 households during the 7 P.M. showing and another 150,000 households during the 2 A.M. showing. A typical *Nova* broadcast reaches between 1.5 and 2 million viewers.

Yet the film was acclaimed by the broadcast industry. In October 1994 it received a British Environment and Media award under the category "Best Network Documentary and Current Affairs," and one year later, in September 1995, it was awarded the prestigious American Emmy award in the category "Outstanding Information or Cultural Program." Of even greater significance, copies of the film were strategically distributed to policymakers and scientists by the World Wildlife Fund and other supporters of the endocrine hypothesis.

The script of the film takes the form of a scientific mystery story, not unlike the early *Nova* broadcasts on public television. One difference between this British film and its public television counterparts in the United States is its intense drama. An opening scene shows a group of scientists stealthily searching the swamps of Florida's Lake Apopka in the dark of night for baby alligators. They are trying to determine the cause of declining alligator populations and the increase in reproductive abnormalities. The audio track—with its pulsating bassoons punctuated by high-pitched, eerie tones—evokes a feeling of danger and mystery.

The film describes discoveries of odd sexual abnormalities in wildlife, including significantly reduced phallus size among some Florida alligators. We are told that not a single Florida panther has normal testes, that male fish in the Great Lakes do not reach maturity, and that there has been a significant increase in reproductive abnormalities among the fish, including the birth of hermaphrodites. Symbols of doom and gloom abound in this documentary. One scientist contributes the dominant metaphor of the film when he declares that "we live, in effect, in a sea of estrogens." After the scenes of swooping helicopters and speedboats cutting through swamp grass in their hunt for alligator nests, we are shown the dramatic effects of feminized wildlife, such as newly hatched alligators without phalluses and with female hormones circulating through their blood. Other species are also affected. Twenty percent of the turtles are said to have an intersex condition, attributed to their exposure to environmental estrogens.

From Florida wildlife, the film segues to the voices of prominent scientists discussing the latest findings on sperm count decline as well as testicular and prostate cancer in humans. The narrator asks: Why are all these things happening at once? Some of the most disturbing footage shows testicular and phallus abnormalities of newborn babies. Scientists speak of high incidences of urethral and testicular diseases in humans while positing a generalized hypothesis that excessive exposure to estrogenic compounds during development could produce these effects.

The documentary showcases British scientists describing controlled experiments with fish placed in the vicinity of sewer outfalls. The experimental fish became hermaphrodites: "Males, in effect, are changing sex," according to the narration. U.K. scientist John Sumpter concludes that "something is coming from the effluent that acts like a female hormone on fish." This finding, we are told, was kept secret from the British public for two years to avoid alarming the population.

In fact, the components of the effluent that were responsible for the sex reversion or ambiguous sexual development were not discovered until late 1996. Initially, the British scientists believed that the estrogenic compound in the sewage had originated from birth control pills. When they were unable to detect the synthetic estradiol in the effluent, they turned their attention to industrial compounds that were known to behave like estrogens, specifically a substance called nonylphenol, widely used in the manufacture of plastics.

Its role as an environmental estrogen had been discovered serendipitously by Ana Soto and Carlos Sonnenschein of the Tufts University School of Medicine. The accidental discovery occurred as Soto and Sonnenschein were seeking to explain the contamination of certain of their control samples, cells that were proliferating without having been intentionally exposed to estrogenic substances. After months of investigation into the possible sources of contamination, the investigators learned that the capped plastic tubes used to store blood serum were shedding an estrogen mimic (a substance that shares some of the functional biological characteristics of an estrogen) that was causing the cells of the control to proliferate. The control thus performed its function of alerting the investigators to a contaminant in their laboratory. The chemical emitted by the plastic tubes was nonylphenol, an antioxidant that gives plastic its flexibility.

After several years of investigation into the endocrine effects of sewage in U.K. rivers, to the surprise of Sumpter and his colleagues, the source of some of the estrogenic chemicals responsible for feminizing male fish was found to be not nonylphenol or other organic pollutants but three hormones linked to natural estrogens found in women's urine, and synthetic estrogens found in birth control pills (see Chapter 3).[17] Yet the British scientists did not rule out other possible sources of estrogenic compounds.

The BBC film ends on a note of existential despair. The narrator states that the endocrine effects of industrial chemicals are real and unknown: "While we still don't know which chemicals or combination of chemicals are causing estrogenic effects or by what route we are being exposed to estrogens, whether through water, diet, or some other means of exposure, the changes to human reproduction are real, and the hunt is still on to pinpoint the key culprit from the sea of estrogens."

It is difficult to predict what impact a prime-time airing of this film would have had on the American public. The images were powerful; the links drawn between wildlife and humans were alarming. The documentary included no skeptics to soften the message of the scientists who spoke of the great urgency to act. A final on-camera statement by a scientist is a call to arms: "Imagine if for the last fifty years we had sprayed the whole earth with nerve gas. Would you be upset? Would I be upset? Yes. I think people would be screaming in the streets. Well, we've done that. We have released endocrine disrupters throughout the world that are having fundamental effects on the immune systems, on the reproductive system. We have good data that show that wildlife and humans are being affected. Should we be upset? Yes, I think we should be fundamentally upset. I think *we* should be screaming in the streets."

Reports of fish abnormalities in 1998 were widely covered in the popular British press and opened new suspicions about industrial pollutants. These studies found that 100 percent of the male roach tested in two rivers in England showed signs of feminization. Large parts of the fishes' sperm-producing testes had turned into egg-making tissue.

The first U.S.-produced hour-long documentary shown on network television that addressed endocrine disrupters was created for WGBH's *Frontline* and titled *Fooling with Nature.* It aired in New England on June 2, 1998, at 9 P.M., with a repeat broadcast on June 7, and was carried on other PBS affiliates throughout the country. Unlike its BBC counterpart, the PBS special treated the issue as a debate among scientists over evidence that humans were at risk. *Frontline*'s producers took exceptional measures to avoid the appearance that the program had a point of view on the issue of whether humans were being adversely affected by the suspect chemicals. This could not be more evident than in the final sequence of the documentary, as viewers hear the distant voices of two contesting scientists, each holding his ground—a symbol of science undecided.

The Nielsen rating for the first PBS airing was 3.0, which translates into a viewing audience of approximately 4.1 million people—nearly eight times greater than the American viewing audience for *Assault on the Male. Frontline* is typically carried on 290 PBS stations, and 30 percent rebroadcast an installment during the week in which it is first aired. Based on survey data compiled by WGBH indicating that 90 percent of those viewing the rebroadcast are new viewers, the weekly cumulative audience for a typical *Frontline* show is estimated at 4.8 million.

Federal agencies were already beginning, albeit slowly, to gain a better understanding of environmental hormone disrupters before *Assault on the Male* was released. At first, congressional actions were uncoordinated and exploratory, but Congress soon directed its concerns about endocrine disrupters into new legislation on food safety and safe drinking water. In March 1996 the first trade book on the subject of endocrine disrupters, *Our Stolen Future,* was released. Although publication of the book was a watershed in presenting arguments in support of the theory of endocrine disrupters, legislative initiatives were already well under way by the time the book reached its intended mass-market readership.

U.S. Legislative Actions

Between 1993 and 1994 eight bills were introduced into the House and Senate that included provisions related to environmental endocrine disrupters. In 1996 the 104th Congress saw six pieces of legislation (three of them associated with amendments to the Safe Drinking Water Act) that addressed the issue of environmental estrogens. For the most part, these bills focused on two themes: research and testing programs.

The Women's Health Environmental Factors Act (H.R. 3509) mandated the Department of Health and Human Services, through the National Institutes of Health, to issue a report on the effects of environmental factors on women's health, specifying that the report should include coverage of "compounds that mimic human estrogens." Another bill, amending Title XIV of the Safe Drinking Water Act (H.R. 3293), added a section titled "Establishment of Screening Program for Estrogenic Substances." Under the bill, the administrator of the EPA, in consultation with the Secretary of Health and Human Services, was required to develop a screening program to determine whether chemicals in commercial use might have effects similar to those of naturally occurring estrogens.

The Omnibus Civilian Science Authorization Act of 1996 (H.R. 3322), passed by the House in September 1996, was the major federal appropriations bill for science. One section, the Endocrine Disrupter Research Planning Act of 1996, required additional coordinated research "to more accurately characterize the risks of endocrine disrupters." Language in the act emphasized the dilemma associated with the environmental endocrine hypothesis, since "neither a conclusion that endocrine disrupters pose an imminent and serious threat to human health and the environment, nor a conclusion that the risks are insignificant or exaggerated, is warranted based on the present state of scientific knowledge." This bill required the EPA to coordinate research in the field of endocrine disrupters, to ensure that risk assessment measures were taken in approving or denying chemical registration, and to see that the research was conducted in a scientific and reasonable manner.

A bill introduced in 1993 and reintroduced in subsequent congresses to amend the Federal Insecticide, Fungicide, and Rodenticide Act (FIFRA) and the Federal Food, Drug and Cosmetic Act (H.R. 1627), known as the Food Quality Protection Act (FQPA), was reported favorably out of the Subcommittee on Health and the Environment and the Committee on Agriculture in July 1996 and signed into law on August 6, 1996. Language in the bill gives the administrator of the EPA authority to "require data or information pertaining to whether the pesticide chemical may have an effect in humans that is similar to an effect produced by naturally occurring estrogen or other endocrine effects."

During the 1995 hearings on the FQPA before the House Subcommittee on Health and the Environment, Erik Olson, senior attorney for the Natural Resources Defense Council (NRDC), one of the nation's most aggressive antipesticide public interest groups, used the DDT experience and evidence from wildlife studies as justification for rewriting the pesticide laws: "Due to pollution from DDT (banned in the 1970s, but still 'out there') and other environmental estrogens, male wildlife are literally being 'feminized': born hermaphroditic with reproductive parts of both sexes or even female entirely." Quoting a Stanford endocrinologist, Olson used the evocative imagery that humans "might be drowning in a sea of estrogens," while emphasizing the additive and synergistic effects of these compounds: "One study was conducted, for example, on the acute toxic interactions of 13 organophosphorus pesticides. The investigators found that 21 pairs had additive toxicity, 18 pairs had less than additive toxicity, and four pairs had synergistic toxicity."[18]

The provisions in the FQPA and the amendments to the Safe Drinking Water Act that call for a screening program for estrogenic chemicals were largely the work of New York Senator Alphonse D'Amato, with support from Senator Daniel Moynihan of the same state. D'Amato's home constituency was Long Island, where breast cancer rates were elevated relative to those in other parts of the state. The activism of Long Island women organized around the issue of breast cancer was a key factor in focusing D'Amato's attention on the issue of estrogenic chemicals and ultimately in winning his support for the screening program. The influential National Breast Cancer Coalition had not played an active role in lobbying for an endocrine disrupter screening program; it focused instead on gaining increased funding for breast cancer research. Now the elevated breast cancer rates on affluent Long Island and in some Massachusetts communities became the stimulus for a new wave of political mobilization among women with the disease.

In 1990 a group of women calling itself One in Nine began lobbying for a comprehensive federal study of breast cancer incidence on Long Island. The group urged the Centers for Disease Control in Atlanta to undertake the study but did not receive a satisfactory response. The leadership of One in Nine spoke to representatives of the National Cancer Institute (NCI) and were told that such a study would cost $5 million. Moreover, institute representatives made it clear that they did not want politics involved in a breast cancer study. Nevertheless, One in Nine President Geri Barish could not imagine raising $5 million without the help of politicians. She approached D'Amato and with members of her board argued the case for the Long Island breast cancer study. D'Amato was responsive to their concerns. According to Barish: "He got us the $5 million from the Department of Defense. . . . After we got the $5 million, the very next day after it was announced that the money was awarded, the NCI pulled out. They said they would not do the study. We called Senator D'Amato and within one hour he called back and said [NCI] is doing the study."[19] The study eventually exceeded the $5 million estimate, and D'Amato found another $15 million to complete the task.

The pursuit of this study was only one part of a larger program of advocacy by One in Nine on behalf of breast cancer victims. Its members lobbied for raising the overall level of research funding for breast cancer and for tighter regulation of pesticides and pesticide residues in food. The Long Island women believed that pesticides might explain the elevated incidence of breast cancer in their region. Federal funding for breast cancer research went from $90 million in 1990 to nearly $500 million five years later in large part because of the political mobilization of breast cancer activists.

D'Amato testified before a joint hearing of the Senate and the House in September 1993, raising the issue of estrogenic chemicals as a possible cause of cancer. He cited the studies of Devra Davis (then at the Department of Health and Human Services), who had worked closely with breast cancer activists, and Mary Wolff at the Mount Sinai Medical Center. It was at these hearings that D'Amato first proposed the idea of testing chemicals for estrogenicity: "Now it appears that there may be a number of environmental agents—pesticides and other chemicals with estrogenic properties—that are capable of increasing a woman's lifetime exposure to estrogen and thus increasing the risk of breast cancer. I believe it is vital that the administration's pesticide bill include provisions requiring routine screening of pesticides for estrogenicity. . . . It is my intent to work on drafting provisions that will mandate estrogenic testing of pesticides."[20]

Two years later, in his statement before the Senate on the FPQA on November 29, 1995, D'Amato revisited the estrogenic testing amendment: "I want to commend and thank the managers of this bill for including in the manager's amendment package our amendment establishing an estrogenic chemical screening program at EPA. . . . This amendment is critical in view of growing evidence linking environmental chemicals that are capable of mimicking or blocking the action of the hormone estrogen to a host of developmental and reproductive abnormalities in wildlife and humans. The most alarming findings suggest a link between exposure to these chemicals and the dramatic increase in human breast cancer that has become so tragically apparent in our Nation over the past several decades."[21]

D'Amato again cited the findings of Mary Wolff of the Mount Sinai Medical Center that higher serum levels of PCBs and DDE were correlated with higher risk of developing breast cancer, suggesting that exposure to estrogenic organochlorines might affect the incidence of hormone-responsive breast cancer. Wolff had testified five months earlier before the House Subcommittee on Health and the Environment that "hormonal disruption may be associated with a broad range of biological effects: reproductive dysfunction, neurological problems, and immunological difficulties" and that endocrine disrupters might be implicated in cancers of the breast, prostate, uterus, ovaries, and colon.[22]

The suspected link between endocrine disrupters and cancer was cited by the Committee on Agriculture in its report on the FQPA: "The Committee is aware of recent scientific reports indicating that some pesticides may initiate, enhance, or block the activity of hormones in humans and wildlife. For example, a linkage has been suggested between human exposure to chemicals that imitate estrogen and breast cancer. Since hormones govern fundamental biological functions such as reproduction, growth, and metabolism in humans and other species, the Committee believes that it is important for EPA to obtain data about the potential hormone-disrupting effects of pesticides in order to make informed regulatory decisions under FIFRA." But the committee did not support a separate amendment to the FQPA calling for the screening of chemicals for endocrine effects. Instead, it said, "The Committee has reviewed and considered the issue and has determined that the EPA currently has sufficient authority to request information related to such effects. . . . Therefore, the Committee expects the Agency, within four years of the date of enactment of this Act, to evaluate the need for and, if necessary, to use its existing authority under sections 3 and 4 of FIFRA to establish standards for data requirements, to determine whether a pesticide can disrupt hormonal activity."[23]

Committee member Congressman George Brown stated his minority view that the endocrine-testing amendment should be included in the bill and highlighted the concern about a breast cancer link to endocrine disrupters: "During the Committee markup I offered, and subsequently withdrew, an amendment to require EPA to develop a data standard for hormonally active pesticides under existing authorities in sections 3 and

4 of FIFRA. These substances, commonly referred to as endocrine disrupters, are believed to interfere with fundamental biological functions such as reproduction and development in humans and other organisms. A link to some types of breast cancer has also been suggested."[24] Brown indicated that he would pursue the statutory amendment on endocrine testing over the committee's negative recommendation when the bill was brought before the House. The House Committee on Commerce, which reported favorably on the FQPA on July 23, included an estrogenic substances screening amendment to the bill. The bill passed the Senate with little opposition.

The screening amendment went further than empowering the EPA administrator to assess endocrine effects; it established a strict timetable: "Not later than 2 years after the date of enactment of this section, the Administrator shall in consultation with the Secretary of Health and Human Services develop a screening program."[25] The screening and testing strategy was to be implemented in the third year and a program report submitted to Congress by the end of the fourth year. This timetable was particularly problematic, because the science for identifying and screening endocrine-modulating chemicals was as yet still in its infancy. The language in the bill put considerable pressure on the EPA to develop a consensus among scientists in an area replete with uncertainty (see Chapter 4).

When the FQPA passed, there were certainly many skeptics within the scientific community who did not believe that ambient levels of industrial and agricultural chemicals were affecting human health by interfering with endocrine functions. The leading journal on the human endocrine system, *Endocrinology,* barely took notice of the hypothesis. Representatives of the industrial sector called the theory that our hormones are under attack a fanciful notion created by environmentalists to bring about a ban on chlorinated compounds and other persistent organic pollutants. Two international protocols in the process of negotiation under the auspices of the United Nations Economic Commission for Europe were prepared through the U.N. convention on long-range transboundary air pollution. The protocols currently cover 16 substances, some of which are known or probable endocrine disrupters.

Yet beyond the initial antienvironmental posturing of certain industrial groups, there was very little opposition to the provision in the new pesticide law requiring screening for endocrine disrupters. Part of the reason for the low level of response in the industrial sector may be found in the issue that had consumed much of their concern over the past decade, namely pesticides and the Delaney clause. First introduced into the Food and Drug Amendments of 1958, the Delaney clause required zero tolerance for food additives that had been shown to be human or animal carcinogens at any dose. The zero-dose standard, however, was never applied to pesticide residues, which could still be found in both fresh and processed food.

The NRDC commenced litigation against the EPA in 1992 on the grounds that Delaney should be applied to pesticide residues, since they are a kind of food additive. Many pesticides then in use would have had to be withdrawn under the Delaney standard. The NRDC's case was eventually upheld by a federal district court, and, through consent decrees based on its court victory, the NRDC was able to increase pesticide prohibitions.

Actions taken by the executive and legislative branches to find a compromise between industry and environmental interests eventually proved successful. Industry spokespersons argued for the replacement of the "zero-risk" standard by a "negligible-risk" standard. Environmentalists, on the other hand, wanted to raise the safety standards for pesticide residues in food, especially for infants and children.

The resulting legislation within the FQPA was one of the least contested environmental acts passed in decades. In the end it was supported by 55 interest groups, including representatives of the agricultural industry, food processors and distributors, and environmental groups. The Committee on Commerce voted 45-0 in support of the FQPA, with the endocrine screening amendment. The bill was passed by a vote of 417-0 in the House on July 23, 1996, and by unanimous consent of the Senate on July 24.

The passage of the FQPA was a victory for stakeholder negotiations as an approach to resolving environmental controversies. Not surprisingly, a few environmental groups (including the Public Interest Research Group and the National Coalition Against the Misuse of Pesticides) opposed the loss of the Delaney zero-risk standard and the emphasis on single pesticide effects, in light of newly publicized reports of chemical synergisms that could increase risk (see Chapter 3). But these broadly supported legislative initiatives on endocrine disrupters are noteworthy for having been initiated well in advance of publication of an influential popular book on the subject, and for not having been prompted by a blockbuster media event or catastrophe, such as the revelation of the toxic waste dump at Love Canal or the chemical spills at Bhopal, India, and Seveso, Italy.

Supported by a small group of stakeholders and influential scientists, the endocrine disrupter screening program was one of the uncontested throw-ins of the new legislation—one that was not perceived to pose any immediate threat to industry. In the context of the major points contested in the FQPA debate, related to pesticides and the Delaney clause, little public attention was directed at a brief provision in the bill titled the "Estrogenic Substances Screening Program." Under the program, the EPA had three years from the date of enactment to determine whether certain substances "may have an effect on humans that is similar to an effect produced by naturally occurring estrogen, or other endocrine effect as the Administrator may designate."[26]

Even if there were any doubts about the public's support for taking a serious look at endocrine disrupters, they would disappear after the publication of *Our Stolen Future* in March 1996.

Our Stolen Future

Theo Colborn had considered writing a popular book to summarize the growing body of evidence in support of her hypothesis that endocrine disrupters deserved serious and immediate attention. She had already drafted chapters that reviewed current laboratory and field studies focusing on the effects of endocrine disrupters on wildlife. In the aftermath of the 1991 Wingspread Work Session, Colborn was continuously seeking bridges to scientists who could advance and bring new insights to the hypothesis. The demands on her time grew rapidly. She fulfilled numerous speaking engagements as word of her scientific findings spread among environmental and industry groups. There were more scientific meetings to attend, consensus statements to draft, and papers to write. It soon became apparent to Colborn that she could not continue networking and building the scientific base for the hypothesis while taking time out to write a popular book. She asked Pete Myers at the W. Alton Jones Foundation if he would collaborate with her on the book. He agreed, but Myers had an intense schedule of his own, and little was gained from this partnership in moving the project forward. Colborn also recognized that her scientific background and writing style might be an impediment to reaching a mass audience. So she began to search for a third collaborator—a professional science writer.

Dianne Dumanoski grew up in central Massachusetts, graduated from Vassar College, and attended graduate school in the mid-1960s at Yale University, where she was enrolled in a doctoral program in English. During the 1968 presidential campaign, before completing her degree, Dumanoski left Yale and worked in public television before beginning a career in print journalism. For over a decade she was a reporter for the *Boston Globe,* where she covered many lead environmental stories, including those dealing with nuclear safety and acid rain.

In December 1992 Dumanoski attended Colborn's invited talk at the Tufts University School of Medicine. After the lecture, Dumanoski interviewed Colborn for a possible story about her work. Sensing the importance of Colborn's thesis, Dumanoski proposed a *Boston Globe* series on endocrine disrupters to her editor, but she was unsuccessful in convincing him to take on the project. She met Colborn, a Pew scholar, again at a meeting of the group, where Dumanoski was asked to engage scientists in a discussion about how to deal with reporters. A month later, while attending a meeting of the Society of Environmental Journalists, Colborn approached Dumanoski with the proposal that she collaborate with Colborn and Myers on a popular book. Although the offer took Dumanoski by surprise, she wasted little time in accepting. Dumanoski described the meeting with Colborn at which the collaboration was first discussed in an essay she wrote for the *Antioch Journal:*

> I assumed that Colborn wanted to tell me about the latest scientific findings in the hope that I would finally write a story. I arrived notebook in hand. As I recall, there was little preamble before she got to the heart of the matter. Would I help her write a book? I was flabbergasted. The proposition seemed preposterous. I didn't know her. She didn't know me. I had never written a book before. I knew a lot about ecological science and the chemistry of ozone depletion, but virtually nothing about the endocrine system. I had only the vaguest notion about how hormones work. . . . I sensed that Colborn was onto evidence that might become one of the most important issues in decades—something as big as ozone depletion. Colborn was likely the next Rachel Carson, but unlike Carson she was not a writer.[27]

Nothing quite prepared Dumanoski for the challenge. She was handed detailed notes in the form of inchoate chapters summarizing an extensive body of scientific studies, observations, and discussions Colborn had had with Canadian scientists. The voluminous body of scientific work and personal reflections needed structure and thematic coherence. The science also needed a social context and a human dimension to engage the general reader. As a writer, Dumanoski was accustomed to being the architect of her own work, building her stories from interviews and information she herself had acquired. Faced with written material from Colborn,

Dumanoski had to decide between starting from scratch and extensively editing Colborn's notes into chapters. She chose to follow both strategies.

In the months that followed, the collaboration among Dumanoski, Colborn, and Myers became highly interactive. They had ongoing discussions about the state of the scientific evidence and how the issues might best be presented to a popular audience. Once the three authors had agreed on a structure for the book, Dumanoski proceeded to supplement the research studies with interviews of scientists and her own investigations into the historical and scholarly literature. Within four months a sample chapter and proposal had been completed and sent out to publishers for bidding. An executive at E. P. Dutton took a special interest in the book based on its prospective marketability, and her company made the three authors a preemptive offer designed to short-circuit any bidding that might take place among publishers. According to Dumanoski, "The advance was substantial for a book promising no celebrity or scandal. It solved our immediate money problems. I now had a salary while I was on leave from my newspaper job, and we had enough for travel and research."[28]

The authors describe the book as an unfolding scientific mystery. In the tradition of such classics as Paul de Kruif's *Microbe Hunters,* Dumanoski assumes the role of narrator and uses the experiences and inquiries of scientists to explore the evidence connecting agricultural and industrial chemicals to hormonal effects on wildlife and humans. Colborn is given a central and well-deserved place in the scientific mystery, taking on the role of master sleuth and serving as the social catalyst who brings about the synergy of the many scientific disciplines that eventually inform the problem. Vice President Albert Gore was approached to provide a quote for use on the book's jacket, but he surprised the authors and their publisher by offering to write the foreword. His foreword compares Colborn's work to that of Rachel Carson in *Silent Spring.* Gore had in fact recently completed the foreword to a new edition of *Silent Spring* published by Houghton Mifflin. In his own book, *Earth in the Balance,* published in 1992, Gore had acknowledged Carson's work but made no mention of Colborn or endocrine disrupters. He did, however, rail against pesticides and the administration of hormones to livestock. His was an invaluable contribution to the public visibility of *Our Stolen Future.*

The most important writing challenge for Dumanoski was to present the science fairly, honestly, and in sufficient detail to make the scientific case without turning off the nonscientist reader. The first hardback edition of *Our Stolen Future* contains 33 pages of endnotes (making up just under 15% of the 250-page narrative). In keeping with its trade format, the book's references support each chapter but are not linked to the text, diminishing its value to scholars but making it less ponderous for the general reader.

The book bases its case for human endocrine effects of chemicals on the wildlife studies and controlled laboratory experiments. Our fate lies with the animals, the authors reiterate. It was a theme that had echoed throughout the pages of *Silent Spring*. The wildlife studies cited in the book were but a small subset of the growing evidence that endocrine disrupters were doing harm. Rather than grounding the hypothesis on one crucial experiment or a single incontrovertible study (there was none), the authors believed that the policy community would be swayed by the weight of the evidence and the diversity of situations it represented. Although there was no dearth of evidence for wildlife effects of endocrine disrupters on which to build an argument, the same could not be said for human effects.

Dumanoski found that there were gaps in Colborn's review of the literature on human health effects of endocrine disrupters. So—with no professional training in biology, toxicology, or environmental health—she began reading the scientific literature, consulting scientists, and tracking down primers on specialized aspects of the human endocrine system so that she could conduct effective interviews in preparation for writing sections on the immune system, cancer, and fertility. Colborn would select the key science articles for Dumanoski to review in preparation for writing a chapter. But the science had to be interpreted and embedded in an interesting story line. Myers—who was both a scientist and a science communicator and who understood the way Colborn viewed the issues—would help frame an interpretation of the science that Colborn could refine and that Dumanoski could then incorporate into a coherent argument. Thus, Myers played a facilitating role between a scientist awash in studies and a science writer seeking a plot that would hold the reader's attention.

Dumanoski spent eight months drafting chapters and sending them to Myers and Colborn for response. The three collaborators met periodically to review the chapter drafts. Some of the book's chapters profiled individual scientists, who were also sent drafts of the relevant manuscript sections for their review. A few of the profiled scientists were opposed to having themselves described in physical, psychological, or personal terms more typical of the biographical genre. The tension between scientific and literary narrative was never more obvious. It fell to Dumanoski to find a balance between these often conflicting styles.

Rachel Carson began *Silent Spring* with a short fable ("A fable for tomorrow"), which described the transformation of a small, bucolic American town into one that had become lifeless—characterized by withered vegetation, vanished birds, dead fish, and infertile farm animals. "This town does not actually exist," she wrote, "but it might easily have a thousand counterparts in America or elsewhere in the world. I know of no community that has experienced all the misfortunes I describe. Yet every one of these disasters has actually happened somewhere, and many real communities have already suffered a substantial number of them."[29] Carson used this literary device to free herself momentarily from gaps in the scientific knowledge of her day, taking license to extrapolate from her observations to an imagined state of environmental despoliation. Similarly, the authors of *Our Stolen Future* introduced one chapter ("To the Ends of the Earth"), combining fact and fiction, that described how PCBs produced in western Massachusetts ended up in the body fat of a polar bear at the Arctic Circle.

Dumanoski explains the importance and the vulnerability of that section: "The polar bear chapter was the first thing I wrote as a sample to accompany the book proposal. I think it is a credit to Pete [Myers] and Theo [Colborn] that they were willing to go along with this kind of approach. They were well aware that scientists would probably grouse about it. I think it signaled their willingness to reach our lay readers. I've gotten more comments from non-scientists about the chapter than anything else in the book."[30] As anticipated, the chapter provided grist for criticism by some scientists (see Chapter 3), who argued that fiction has no place in a serious, scientifically grounded account of environmental hazards. The criticism echoed back to that of the fable in *Silent Spring.* Carson biographer Linda Lear noted: "The fable was almost uniformly derided by reviewers unable to understand its basis in allegory and used it to further demean her credibility as a scientist."[31] Yet by other accounts, the PCB fable was a plausible, scientifically credible scenario that was well received by general readers and brought home the point that environmentalist Barry Commoner had made in his 1971 classic *The Closing Circle,* in what he called "The First Law of Ecology": the entire biosphere is interconnected.

As we have seen, government policymakers were already beginning to take notice of Colborn's findings several years before the release of *Our Stolen Future.* Agency scientists were invited to key meetings and reported back to their colleagues. Congressional hearings quickly spread word of the issues through Washington policy circles. European countries were also showing grave concern over organochlorines—concern that made them willing to take a serious look at endocrine disrupters.

The book's publication, then, was one of the catalysts responsible for coalescing interests over the role of persistent organic chemicals in the environment. By 1999, 16 foreign editions had been published, including British, French, Spanish, Portuguese, Brazilian, German, Polish, Slovenian, Dutch (for serial publication in a magazine), Swedish, Danish, Norwegian, Korean, and Japanese versions, as well as two Chinese editions, one for the People's Republic of China and a separate Taiwanese volume. By early 1999 the Japanese publisher alone had printed 173,000 and sold over 120,000 copies of its edition, and by January 31, 1999, U.S. sales of *Our Stolen Future* had surpassed 62,000 copies. Global awareness of endocrine disrupters had begun to take root. In this climate of growing worldwide awareness, U.S. regulatory agencies were poised to take action, with leadership coming from no less a source than the Executive Office of the President.

Executive Branch Initiatives

By the mid-1990s endocrine disrupters had caught the attention of the White House. Vice President Gore's knowledge of and involvement in global environmental concerns made him receptive to the subject. He seemed to have become apprised of the problem sometime after the completion of his 1992 book *Earth in the Balance,* and there is some indication that he had had an opportunity to view the BBC documentary *Assault on the Male.* He certainly also had access to a prepublication copy of *Our Stolen Future* in preparation for writing the foreword to the book.

In his foreword, Gore describes Colborn's efforts in sounding the alarm on synthetic chemicals as the sequel to the work of Rachel Carson: "*Our Stolen Future* takes up where Carson left off and reviews a large and growing body of scientific evidence linking synthetic chemicals to aberrant sexual development and behavioral and reproductive problems. Although much of the evidence these scientific studies review is for animal populations and ecological effects, there are important implications for human health as well."[32] It was a comparison that environmental activists found energizing and environmental conservatives viewed as blatantly premature.

During President Bill Clinton's reorganization of the Executive Office, he had established by executive order in November 1993 the National Science and Technology Council (NSTC), a cabinet-level committee whose function was to set priorities for and to coordinate approaches to science and technology issues across the federal government. Prior to Clinton, U.S. presidents had been advised on science policy issues by a science adviser, who directed the Executive Office of Science and Technology Policy. The new cabinet-level body elevated the importance of science and technology in the Clinton administration.

In principle, the president chairs the NSTC, which also consists of cabinet-level administrators, the vice president, and the president's chief science adviser. The NSTC operates through nine committees, each chaired by a senior agency official and co-chaired by a member of the Office of Science and Technology Policy. The various committees of the NSTC advise the president on research priorities and budgeting decisions that reflect those priorities. The director of the Office of Management and Budget sits as a member of the NSTC, all of whose committees have a designated liaison within the budget office. Priorities selected by the NSTC, in theory, receive the imprimatur of the president. This administrative structure is intended to provide top-down coordination from the Office of the President through all agency heads.

One of the committees of the NSTC, the Committee on Environment and Natural Resources (CENR), had been formed to coordinate the work of an interagency group of 26 government scientists representing 14 federal agencies. In November 1995 the CENR identified five national research priorities: global climate change, natural disasters, environmental monitoring, a North American strategy for reducing tropospheric ozone, and endocrine disrupters. With regard to endocrine disrupters, a working group of the CENR identified four objectives for developing an integrated research strategy across the federal agencies: a planning framework for federal research on the human health and ecological effects of endocrine disrupters, an inventory of ongoing federally funded research, identification of research gaps, and the development of a coordinated interagency research plan. By November 1996 the committee's working group had produced the planning framework, titled *The Health and Ecological Effects of Endocrine-Disrupting Chemicals: A Framework for Planning.* During the summer of 1996, a White House task force began working on the inventory of all federally funded endocrine disrupter–related research. The finished inventory described nearly 400 projects, more than half of which were focused on PCBs (109) and dioxins (87). In addition, there were 41 projects on DDT and DDE, 25 on phytoestrogens, 18 on organochlorines, and 11 on oral contraceptives.

In 1997 federal budgetary allocations for the five NSTC priorities were approximately $1.9 billion for global climate change, $1 billion for natural hazards, $600 million for environmental monitoring, $150 million for tropospheric ozone reduction, and about $30 to $50 million for endocrine disrupters. (Since the funding for endocrine disrupters was not a line item in the budget—unlike, for example, global climate change—the budget estimate for research included many areas of funding not explicitly allocated to solve the problems raised by the environmental endocrine hypothesis, but that were nevertheless considered relevant in compiling the agency inventories.)

According to *Environmental Science & Technology*, White House officials estimated that the federal government spent about $30 million annually on research directly focused on endocrine disrupters, but they noted that the amount could easily be as high as $500 million if research that indirectly examined endocrine disruption were figured in.[33] An inventory of endocrine disrupter research projects such as that developed by the White House task force provided a new organizing schema for classifying existing research. In reality, only a small percentage of the estimated total research funds for endocrine disrupters represented new initiatives. Yet the emphasis on reproductive and developmental toxicity, viewed within the context of endocrine disrupters, allowed policymakers to classify the disparate research programs as reflecting a single theme.

The National Institute of Environmental Health Sciences (NIEHS) had been the federal agency first on the scene in considering the potential problems of endocrine disrupters. The visibility of its effort was due largely to John McLachlan, who had spent 21 years with the agency in its Laboratory of Reproductive and Developmental Toxicology and who headed its Division of Intramural Research between 1988 and 1994. In 1987 McLachlan coauthored a paper in *Environmental Health Perspectives* (the official journal of the NIEHS) titled "Estrogens and Development," in which he postulated potential effects on the development of mammals resulting from exposure to environmental estrogens. The abstract of his paper stated: "The normal development of the genital organs of mammals, including humans, is under hormonal control. A role for the female sex hormone estrogen in this process is still unclear. However, exposure of experimental animals or humans to the potent exogenous estrogen, diethylstilbestrol (DES), results in persistent differentiation effects. Since many chemicals in the environment are weakly estrogenic, the possibility of hormonally altered differentiation must be considered."[34]

McLachlan's thesis about exogenous estrogens grew out of his studies of DES and preceded the wildlife studies that had caught Theo Colborn's interest when she began her work for the Conservation Foundation in 1987. McLachlan and his collaborator found in their DES-treated mice many of the abnormalities reported in humans exposed prenatally to DES, including structural malformations in the uterus, multiple abnormalities in the reproductive tract, malformed oocytes, and clear cell carcinoma.[35] The strong correlation between the effects of this potent xenoestrogen in mice and humans established the credibility of the mouse model for studying chemical estrogen mimics and antagonists.

In 1995–96, the NIEHS issued a request for proposals under the initiative "Environment and Women's Health." The program called for research into the role of environmental estrogens in causing cancer, endometriosis, osteoporosis, benign tumors, reproductive disorders, and neuroendocrine and autoimmune disorders. The NIEH's request stated that cancer incidence from DES exposure might not have peaked and asked for continued epidemiological studies of DES-exposed patients as well as basic research using DES as a model for other estrogenic chemicals.

The Environmental Protection Agency is an action agency—one that turns science into policy. In its strategic plan for the 1990s, endocrine disrupters were one of six research priorities, along with ecosystem protection, drinking water disinfection, human health protection, particulate matter reduction, and pollution prevention. Citing evidence from wildlife exposed to high levels of organochlorines, complex sewage and industrial effluents, and natural sources such as phytoestrogens, and following a series of workshops, in 1996 the EPA consolidated its existing research initiatives in reproductive toxicology under the Endocrine Disrupter Research Program. The stated goal of the program was to evaluate the potential health effects associated with endocrine disrupters and to determine the extent of current exposures. The EPA defined an endocrine disrupter as any exogenous agent that interferes with the production, release, transport, metabolism, binding action, or elimination of natural hormones in the body responsible for the maintenance of homeostasis and the regulation of reproductive and developmental processes. The agency sought investigator-initiated proposals in four broad areas: human health effects, ecological health effects, human exposure evaluations, and ecological exposure evaluations of endocrine disrupters. For fiscal year 1996, the EPA allocated $3.5 million for its Endocrine Disrupter Research Program.

The EPA action was a response to the NSTC's National Research Strategy on Endocrine Disrupting Chemicals. At a workshop titled

"Research Needs for the Risk Assessment of Health and Environmental Effects of Endocrine Disrupters," the EPA established its own research priorities. The report of the workshop was published in *Environmental Health Perspectives,* and the participants' interpretation of the evidence for the endocrine hypothesis was expressed in the introduction to the workshop statement:

> Evidence has been accumulating which indicates that humans and domestic and wildlife species have suffered adverse health consequences from exposure to environmental chemicals that interact with the endocrine system. To date, these problems have been identified primarily in domestic or wildlife species with relatively high exposures to organochlorine compounds, including [DDT] and its metabolites, PCBs and dioxins, or to naturally occurring plant estrogens. It is not known if similar effects are occurring in the general human population, but again there is evidence of adverse effects in populations with relatively high exposures. . . . The critical issue is whether sufficiently high levels of endocrine-disrupting chemicals exist in the ambient environment to exert adverse health effects in the general population.[36]

The workshop participants concluded that future research should focus on evaluating the validity of the hypothesis, the effects of endocrine disrupters on reproductive development, improved exposure assessment, and the effects of chemical mixtures. The EPA also published proposed updated testing guidelines for evaluating developmental and reproductive effects of pesticides and industrial chemicals. The guidelines specified the agency's intention to develop a screen for endocrine disrupters.

The EPA recently revised its guidelines for developmental and reproductive toxicity testing and recommended testing for sensitive endocrine end points, such as sperm count, quality, and activity, and hormonal fluctuations in the female reproductive cycle. In a June 1996 report the EPA identified six uncertainties it faced in regulating endocrine disrupters.

1. What classes of chemicals may affect the endocrine system?
2. How much exposure to these chemicals does it take to produce adverse effects?
3. How are humans and wildlife exposed?
4. What are the combined effects of exposure to multiple endocrine disrupters?

5. What effects are actually occurring among exposed humans and wildlife?
6. Do current guidelines for chemical testing adequately predict these effects?

While collaborating with other agencies and funding its own research in pursuit of the answers to these questions, the EPA was also reevaluating its regulatory agenda in terms of endocrine effects. It cited four organochlorine pesticides still on the market that act as endocrine disrupters—dicofol, methoxychlor, lindane, and endosulfan—of which it was undertaking a reassessment. In 1996 the agency was far from the point where it could justify removing a chemical from the market exclusively on the grounds that it was an endocrine disrupter. Nevertheless, it began to look at suspect chemicals with endocrine disruption as another indicator of concern. In 1997 a technical panel of EPA scientists issued a report summarizing the current state of knowledge of endocrine-modifying chemicals. Curiously, the agency had contracted for a similar analysis (perhaps with a little more attention paid to weaknesses in the available knowledge) from the NAS. In an era of cost containment and efforts to hone regulatory efficiency, one might have expected more thorough pooling of scientific resources within the government.

The EPA and the Department of Interior had contracted with the NAS Board on Environmental Studies and Toxicology for a panel study that would critically review the literature on hormone-related toxicants in the environment. The cost of the study was set at $860,000, with the EPA contributing $220,000 and Interior paying $640,000. Among the stated goals of the study were: "to review critically the literature on hormone-related toxicants in the environment; identify the known and suspected toxicological mechanisms and impacts on fish, wildlife, and humans; identify significant uncertainties, limitations of knowledge, and weaknesses in the available evidence; develop a science-based conceptual framework for assessing observed phenomena; and recommend research, monitoring, and testing priorities."[37] The NAS study was expected to provide a consesus position on the state of knowledge of endocrine disrupters and to recommend research, monitoring, and testing priorities. In addition, the panel's report was supposed to identify particular chemical substances with hormone-receptor binding properties and to characterize known and potential problems posed by endocrine disrupters.

The Board on Environmental Studies and Toxicology established a study panel called the Committee on Hormone-Related Toxicants in the Environment. The structure established by the NAS and its research arm, the National Research Council (NRC), for this type of study called for multiple tiers of review: the working committee was to write and approve a draft report; an external review panel would examine the draft report and recommend changes; the NRC staff and the working committee would respond to the draft; approval would then be given by the coordinator of the working committee; the report would be reviewed by liaisons to NAS/NRC within the academy's Commission on Life Sciences; and final approval would be granted by the NAS/NRC executive director of life sciences.

Seventeen scientists made up the working committee. Some of them had previously expressed viewpoints on endocrine disrupters, whereas others were not as yet associated with a public position on the issue. Stephen Safe, professor of veterinary physiology and pharmacology at Texas A&M University, was the most publicly visible skeptic of the claim that there is an effect on human health from xenobiotics. Louis Guillette of the University of Florida, on the other hand, was one of several committee members who had taken public positions in support of the hypothesis. The endocrine disrupter committee began its work in October 1995 and after five meetings was expected to arrive at a consensus document scheduled for release in late 1997. However, after dozens of unsuccessful attempts to arrive at a unified position, the consensus proved far more complicated than expected. By late August 1998 the NAS/NRC draft report, complete with a minority statement, was released to external reviewers. The external reviews were more supportive of parts of the minority position than some had expected. In response, the NRC's oversight group for the panel study asked several members of the committee whom they viewed as moderates on the endocrine disrupter hypothesis to make a last effort at developing a unified consensus position. To avoid adding a minority report, the committee agreed in principle to prepare a single document that would include sections describing the areas in which consensus could not be reached. There were factional disagreements over whether human fertility was in decline, what studies should be included and emphasized in the report, the cause(s) of developmental abnormalities in some wildlife species, and the degree to which humans were at risk from environmental levels of endocrine disrupters.

With the study caught up in these internecine conflicts, the release date for its report was extended into 1999, nearly two years after its stipulated date of completion. By May 1999 there were still residual conflicts, but it appeared that the academy report was leaning toward the following conclusions: The committee could not say with confidence whether endocrine disrupters in the environment posed a major threat to the general population that required an immediate national response. However, the potential threat of endocrine disrupters should not be ignored, since these substances are known to have adversely affected wildlife populations, and laboratory studies have demonstrated that they can affect developmental and reproductive processes in vertebrates. The committee was also in general agreement on the need for vigilant monitoring of wildlife and human populations and further research into the mechanisms of xenobiotic endocrine effects.

It should come as no surprise to careful observers of the NAS that separation of science from values and politics, even in the nation's preeminent scientific body, is the exception rather than the rule. The final academy report on endocrine disrupters (which it euphemistically refers to as "hormonally active agents") will most certainly give the imprimatur of the academy to the importance of these chemicals as the actual cause for some (and the potential cause for other) human and wildlife health effects. In addition, the academy study will most assuredly legitimate some of the concerns raised by the dozens of scientific signatories to the Wingspread consensus statements.

The EPA, meanwhile, did not wait for the release of the NAS panel study to report its own results, thus sending a confusing message to the public and appearing to short-circuit the role of the NAS.[38] Agency scientists examined nearly 300 peer-reviewed studies involving laboratory animals, humans, and wildlife. There were no surprises in their results. The report indicated that, although some animal effects of endocrine disrupters had been demonstrated, except for a few unusual circumstances the authors could not draw causal connections between endocrine disrupters and human disease or abnormalities. The media response to the EPA report was to conclude that the human health effects of endocrine disrupters are uncertain and in need of further study. Some observers considered the report an exercise in artful ambiguity. For example, the signatories to the EPA report used the term *working hypothesis* in referring to the theory of endocrine-disrupting chemicals. "With a few exceptions . . . a causal relationship between exposure to a specific environmental agent and an adverse health effect in humans operating via endocrine disruption has not been established."[39] The exceptions cited by the report included workplace exposures and DES. It emphasized that the evidence linking environmental exposure to endocrine disrupters with infertility or cancers of the breast or prostate was not conclusive.

The inconclusive nature of the EPA report and the delay in release of the NAS findings notwithstanding, other federal agencies had reclassified program funds in response to the new concerns about the hormonal effects of environmental chemicals. The NCI Joint Research Project consists of three agencies (NCI, EPA, and NIEHS) that have pooled resources to evaluate the role of agricultural pesticide exposures in cancer, neurologic disorders, reproductive abnormalities, childhood developmental problems, and other chronic diseases. One study has been following 112,000 adults and their children in farming communities in North Carolina and Iowa to assess the role of diet and lifestyle factors in the development of one or more of these conditions. The U.S. Fish and Wildlife Service initiated a number of research projects assessing the impacts of PCBs, dioxin, organochlorines, and metals on the long-term survivorship and fertility of raptors, seabirds, ducks, and fish. The effect of endocrine disrupters on wildlife had not created a public outcry, but slowly and deliberately the message of *Our Stolen Future* that wildlife served as biological sentinels and precursors of human chemical hazards was getting through to policymakers, even as the human data were still in their infancy. Meanwhile, beyond the United States, the European community and several of the industrialized Asian nations were also responding to the new threat of environmental endocrine disrupters.

International Activities

The International Organization for Economic Cooperation and Development, most commonly referred to by its acronym OECD, was founded in 1961 with 20 member states. It currently has 29 members from Asia, Europe, and the Americas, including the United States. Its newest member is the Czech Republic, which joined in 1995.

The OECD has an Environmental Health and Safety Division that focuses on the regulatory management of chemicals, publishing guidelines for the testing, risk assessment, and handling of chemical agents. Its scope includes pesticides, food additives, pharmaceuticals, and the products of modern biotechnology. In 1996 the division initiated an accelerated response to endocrine disrupters. In September of that year, the national coordinators of the OECD's Test Guidelines Program proposed that something be done about incorporating endocrine disrupters into the organization's chemical testing protocols, by either issuing new guidelines or revising existing ones. Meanwhile, the OECD began collaborating with other international organizations. Working with the European

Union and the World Health Organization, it organized a workshop on endocrine disrupters in Weybridge, England, in December 1996. From that meeting came working definitions of actual and potential endocrine disrupters. Workshop participants defined an endocrine-disrupting chemical as "any exogenous substance that causes adverse health effects in an intact organism, or its progeny, consequent to changes in endocrine function." A potential endocrine disrupter was defined as a substance that "possesses properties that might be expected to lead to endocrine disruption in an intact organism."[40] Both definitions emphasize the critical role of in vivo studies in defining endocrine disrupters. Biochemical results are suggestive, but they cannot, in terms of the working definitions, be considered a sufficient criterion to classify a chemical as an endocrine disrupter in the absence of studies done in whole animals. This approach is consistent with the line of thinking that the process of testing chemicals should involve a tiered series of screens, perhaps beginning with biochemical assays but certainly ending with whole-animal studies. According to the Weybridge definitions, chemicals that are found to bind to hormone receptors would not be classified as endocrine disrupters unless and until it could be shown that the hormone mimics cause adverse health effects—a standard some would find too high to use in developing policy.

Other U.S.-European collaborations were organized to harmonize international activities and to avoid duplication of effort. In November 1996 the White House's CENR sponsored the first in a series of workshops with the United Nations Environmental Program, the EPA, and the OECD. Another international collaboration began in 1997 among the EPA, the OECD, and the International Program in Chemical Safety of the World Health Organization. Among the goals of this collaboration were the development of an international assessment of the state of the science of endocrine disrupters and a global inventory of research activities. The collaboration was expected eventually to result in a report titled *International Assessment of the State of the [Endocrine Disrupter] Science Report.*

Anyone reviewing the scientific literature on endocrine disrupters soon comes to appreciate its multidisciplinary and fragmented nature. Biomedical scientists rarely consult more than half a dozen journals on an ongoing basis. Investigations into the potential causal relationships between chemical exposures and reproductive and developmental effects in humans may require many types of reinforcing studies to build the argument from the cellular and molecular levels to occupational health and epidemiological studies. The standard electronic databases such as Medline are often too narrow to cover the spectrum of studies that could provide valuable knowledge about the environmental effects of endocrine disrupters. The international collaborations moved rapidly to develop inventories of research activities that crossed traditional sector boundaries. The EPA had taken an early initiative in developing a national inventory of research projects on endocrine disrupters. The U.S.-European collaboration was intended to expand the EPA inventory into a global database of endocrine disrupter research classified by government, industry, and nongovernmental organizations (NGOs), with Internet access.

The years 1996 and 1997 witnessed a significant expansion of international activities on endocrine disrupters. Ecological and human health issues defined by chemical endocrine disrupters (sometimes also referred to as *environmental toxicants* or, to use industry's preferred terms, *endocrine modulators* and *endocrine-active substances*) provided a new framework and organizing principle for identifying connections among disparate forms of research on animal and human health issues. The boundaries of science were being recast. For the first time scientists involved in studies of fertility, transplacental movement of chemicals, and thyroid disorders in wildlife—subjects related within the broad outlines of the endocrine system—found themselves listed in the same research inventory, realizing that they perhaps had reasons to read each other's papers and communicate with one another. By March 1997 the European Endocrine Research Inventory contained 89 topics related to endocrine disrupters affecting humans and wildlife.

The OECD conducted a survey of its member states between December 1996 and May 1997 to obtain a snapshot of country activities and perspectives related to endocrine disrupters. The survey was completed by 21 of its 29 member states. At least half the respondents considered endocrine disrupters a major level of concern. But except for the United States, no country indicated that regulatory actions had been taken or were being prepared specifically to address endocrine disrupters—although all the respondents considered these chemicals candidates for future regulatory or advisory activities. Most of the respondents believed that the OECD's current guidelines were insufficient to address the problem. The majority of countries reported that they already took part in one or more international activities that addressed the management of risk from endocrine disrupters, such as the World Health Organization, the European Union, or the United Nations Environmental Program. The OECD was identified as the international body expected to play the lead role in developing screening and testing protocols for endocrine disrupters.

In April 1997, the OECD participated in a European expert workshop, "Endocrine Modulators and Wildlife: Assessment and Testing." A three-tiered testing strategy was recommended, building up from an initial assessment, to developing screening priorities, and finally to using in vivo tests to determine hazardous concentrations or doses of endocrine disrupters. By the spring of 1998, the OECD could report that its member countries had reached a consensus on an approach to the identification and assessment of endocrine disrupters. The first meeting of the OECD Endocrine Disrupter Testing and Assessment Working Group was held in Paris in March 1998. Its goal, like that of its U.S. counterpart within the EPA, was to develop methods for testing the effects of endocrine disrupters and a program for validating the tests. In late 1998, the European Parliament voted 497–4 for a gradual withdrawal of hormone-disrupting chemicals from the European market—even before a generally accepted inventory of harmful substances had been finalized.

When Japanese authorities reported that male sperm counts had fallen 10 percent from the levels of the 1970s and that hormone-disrupting chemicals had been detected in harbors and rivers, their government proposed a $10 billion investment in endocrine disrupter research. The proposal included a research center within the National Institute for Environmental Studies to be devoted to studying environmental endocrine effects. Japanese scientists also formed the Japan Endocrine Disrupters Society to help shape the research agenda of the new government initiatives. Japan held its first international conference on endocrine disrupters in December 1998.

It is a credit to a quarter-century of environmental activism that the regulatory wheels began moving within only a few years after the first Wingspread workshop. But veteran observers of government know that *sustained* activity, particularly as congressional fashions change, requires strong and continuous external constituencies leaning on Congress and federal agencies. The independent sectors consisting of NGOs play at least two roles in shaping policy: they act as interest groups that seek to influence policy by working directly with congressional staff and agency personnel, and they serve as communication links to the broader society, networking and building constituencies for their agendas. With the expanded role of the Internet, public interest groups and industry-sponsored organizations were beginning to capitalize on the new free-form global communication strategies to shape world opinion on endocrine disrupters.

As the concerns over endocrine disrupters spread throughout the world community, NGOs, both new and established, began responding.

Some viewed the situation as an opportunity to tighten their countries' environmental regulations. Others, supported by friends of the chemical industry, responded by defending chemical products suspected of having hormonal impacts.

The Role of Nongovernmental Organizations

Ordinarily, NGOs play a downstream role with respect to the knowledge base that drives environmental policy: most public interest groups depend on governmental and academic sources for the information needed to support their advocacy work. But NGOs are increasingly becoming involved in their own research. The Natural Resources Defense Council selected its own scientists to evaluate the cancer risks to children of pesticide residues in food. As a result of their study and the attention it received in the media, the public created a self-imposed ban on Alar, which eventually drove the pesticide off the market. Some organizations use scientific data from the federal government but provide their own syntheses and interpretations of the information in reports to the public. As an example, in January 1998 the Environmental Working Group released a study on pesticides and children's food that describes the hazards, particularly to young children, of organophosphate residues.

Although endocrine disrupters, known at first by other designations, have been studied for over 20 years through federally funded research, it was the support of public interest NGOs like the Conservation Foundation, the World Wildlife Fund, and the W. Alton Jones Foundation that made it possible for Theo Colborn's formulation of the environmental endocrine hypothesis to be brought to the attention of a vast number of scientists and policymakers. Other NGOs, created or supported by industrial organizations, opposed the rising tide of environmental groups and media analyses by responding to claims that threatened industrial products. A brief accounting of the burgeoning NGO activities focused on endocrine disrupters during the period 1995–97 follows. The explosive growth of this issue within such a short time without a signature event or definitive evidence of human illness is quite unusual in modern environmental history.

Silent Spring Institute. Founded in 1994, the Silent Spring Institute is a nonprofit scientific research organization that investigates links between the environment and diseases that affect women. In 1993, Massachusetts, one of several states that maintains a cancer registry, released data indicating that breast cancer incidence for postmenopausal women was significantly elevated compared with the statewide average in 9 of the 15 towns on Cape Cod. The institute was awarded a contract by the state of Massachusetts to carry out what came to be called the Cape Cod Breast Cancer and Environment Study, the objective of which was to find patterns of association between breast cancer and environmental factors.

The study included a component that examined the possibility that endocrine disrupters in drinking water were a risk factor for breast cancer. In this region of Massachusetts, endocrine disrupters were suspected as possible causal agents because of extensive aerial applications of pesticides to cranberry bogs, use of chemicals for turf management on golf courses, residential solvent use, and the presence of contaminant plumes from sites such as the Massachusetts Military Reservation. The study marked the first effort to use statewide cancer incidence data to explore the human health effects of environmental endocrine disrupters.

The Silent Spring Institute was also involved in community education among the residents of Cape Cod. It distributed a community information bulletin explaining the meaning of hormone-disrupting chemicals and the reasons they might be linked to breast cancer. In its statement outlining the proposed study, the institute wrote: "A third factor pointing us in the direction of [environmental factors] is that wastewater, which is in general an important source of drinking water contamination, is known to be estrogenic, and estrogen is generally considered to be a strong risk factor for breast cancer. Recent research has shown that a number of chemicals, including pesticides, can act like estrogens. These findings have led to the hypothesis that exposure to these compounds could potentially increase risk of breast cancer."[41]

The institute contracted with Tufts University School of Medicine professors Ana Soto and Carlos Sonnenschein, developers of the E-SCREEN assay (which uses human breast cells in a culture medium to measure estrogenicity),[42] to apply the assay to samples of waste water and public and private drinking water. The investigators and the institute were able to advance the state of the art of xenoestrogen analysis by perfecting extraction methods for preparing samples for the E-SCREEN assay. Since human dose-responses to xenoestrogens are not known, the institute was faced with the need to apply the E-SCREEN to trace amounts of chemicals that had to be carefully extracted from environmental samples. The Tufts scientists perfected methods of measuring *estrogenic equivalents* that would enable one to make comparisons among different chemical mixtures and to evaluate the potency of xenoestrogen "cocktails." Using these methods they were able to detect estrogenic activity in chemical mixtures in wastewater and drinking water samples without having to know which of the chemicals were active estrogenic compounds.

The first phase of the Cape Cod cancer study did not solve the mystery of the elevated breast cancer incidence in the region. The E-SCREEN analysis found high levels of endocrine disrupters in wastewater and moderate levels in aquifers, but based on the limited samples available, the levels in the Cape's drinking water were not found to be especially high. At least in the early phases of the study, they could not account for the elevated cancer rates in the region.

Professional Scientific Societies. Professional scientific societies have been slow to respond to the environmental endocrine hypothesis—despite the fact that endocrine disrupters have gone from being the subject of simple poster presentations, which typically receive very little attention from the scientific mainstream, to being the centerpiece of entire symposia. Nevertheless, several professional societies began to take a serious interest in the issue after the announcement of the NAS study and the expansion of new funding opportunities.

The Society for the Study of Reproduction began with 1 poster session in 1993, expanding to 5 sessions in 1994 and 12 sessions in 1995, along with a minisymposium titled "Disruption of Reproduction Function by Environmentally Relevant Chemicals: Novel Mechanisms." In 1993 the discussion was limited to dioxin, but by 1995 it included organochlorine pesticides and naturally occurring estrogenic compounds.

The Society for Environmental Toxicology and Chemistry (both the U.S. organization, SETAC, and its European counterpart, SETAC Europe) has been monitoring the endocrine-disrupting effects of dioxin for a number of years. At its 1995 meeting, the society focused entirely on the reproductive and developmental effects of dioxin. In 1997, the group organized its first day-long symposium on endocrine disrupters, and SETAC Europe also held a meeting on the subject.

The American Association of Zoologists (now the Society for Integrative and Comparative Biology) featured a symposium on endocrine-disrupting chemical effects on wildlife populations during its 1995 and 1998 meetings. The Conference on Great Lakes Research continued to draw attention to endocrine disrupters by holding a symposium on the topic during its 1996 meeting, held at the University of Toronto. The International Neurotoxicology Conference, held October 29 through November 1, 1995, in

Hot Springs, Arkansas, titled its meeting "Developmental Neurotoxicology of Endocrine Disrupters: Dioxin, PCBs, Pesticides, Metals, Psychoactive and Therapeutic Drugs," addressing evidence for human and wildlife effects, as well as ways to improve techniques for risk assessment. The proceedings of these meetings are generally published in the field's journal of record, *Neurotoxicology.*

For two years Chris Wilkinson, a scientist with a private consulting group, was unsuccessful in convincing the program committee of his scientific society, the International Society of Regulatory Toxicology and Pharmacology, to hold a meeting on endocrine disrupters. He finally succeeded for the 1997 meeting. The annual meeting typically focuses on a single theme, and that year it was "Assessing the Risks of Adverse Endocrine-Mediated Effects." Following the format of toxicological risk assessment, the sections were divided into considerations of dose-response issues, human exposure to endocrine-active chemicals, testing for endocrine-mediated effects, and endocrine risk characterization. Endocrinologists studying humans were slow to respond to news of the possible effects of environmental hormone disrupters. Very few published articles appeared in the leading endocrinology journals during the early 1990s. But by 1997 there were signs of change, and the joint meeting of the Endocrine Society and the International Society of Endocrinology held two sessions on environmental influences on hormones.

The circle of interest and participation among scientific and medical fields and private sector organizations was widening. It soon extended beyond scientific societies. In 1996 the International Business Communications Conference was devoted to endocrine disrupters. The corporate world understood the importance of keeping up with the learning curve in an area in which the regulatory climate was so fragile. Public interest and advocacy groups were quick to incorporate such concerns about endocrine disrupters into their own environmental agendas.

Greenpeace International. Greenpeace has been calling for a global ban on chlorinated compounds based on the intergenerational effects of those chemicals. The group has argued that current restrictions based on individual chemicals can, at best, only hope to stabilize global environmental levels, allowing toxins to continue concentrating in the tissues of humans and wildlife. Greenpeace's global strategy has included advancement of the "precautionary principle" (which states that certain activities or the manufacture of certain products should cease even when there are only hypothetical and untested risks), a global treaty banning organochlorines, a progressive tax on chlorine production, a transition fund to aid displaced workers, and technical and financial assistance for developing countries and small businesses that rely on organochlorines. In its campaign against chlorinated chemicals, Greenpeace emphasized that "some organochlorines may cause damage by either mimicking or blocking steroid hormones which affect growth and sex, in particular the female sex hormone oestrogen and the male sex hormone testosterone."[43]

W. Alton Jones Foundation. The W. Alton Jones Foundation is the leading private foundation supporting studies and educational programs on endocrine disrupters. Under its Sustainable World Program, the foundation dedicated substantial funding for projects seeking to eliminate contamination affecting children's health, with special consideration given to pesticides and endocrine disrupters. It awarded over $1.1 million in 1994 and over $1.8 million in 1995 to research, public education, and grass roots activities directed at understanding and reducing the use of endocrine disrupters and pesticides. Under the directorship of John Peterson Myers, the foundation was an early supporter of the work of Theo Colborn while she was a senior fellow at the foundation and at the World Wildlife Fund. Research supported by the foundation has included surveys by Tulane University researchers of coastal Louisiana ecosystems to assess the impact of endocrine disrupters on wildlife and human health. The foundation also contributed to several of the Wingspread Work Sessions.

World Wildlife Fund. The fund has taken a strong stance against organochlorines and endocrine disrupters since wildlife evidence began accumulating around the Great Lakes. By supporting the work of its senior fellow (and later senior scientist) Theo Colborn in investigating the effects of endocrine disrupters, the fund has positioned itself for advancing not only the science but also the communication of risks and the formulation of public policy regarding this class of chemicals. The organization has embraced the environmental endocrine hypothesis. For example, in the summer 1995 issue of its publication *Eagle Eye,* WWF-Canada stated that "some synthetic compounds have now been shown to disrupt endocrine function in the body." A video produced by WWF-Canada titled *Hormone Copy-Cats* brought Theo Colborn's message to a broader audience of activists, educators, and policymakers.

Advancement of Sound Science Coalition. The coalition is a Washington-based, nonprofit, corporate-supported organization dedicated to "ensuring the use of sound science in public policy decisions." Describing the group, the Multinational Monitor (a progressive corporate watchdog organization) distinguished between its stated aim (to combat the "consequences of inappropriate science through focusing attention on current examples of unsound government research used to guide policy decisions") and its real agenda ("to oppose any safety or health regulations that might impinge on their bottom line").[44] A press statement following the publication of *Our Stolen Future* characterized the coalition's approach: "'Partial, incomplete science that serves only to alarm the public with fears, not facts,' is the way Garrey Carruthers [former Governor of New Mexico], chairman of the Advancement of Sound Science Coalition, described *Our Stolen Future,* a new book receiving heavy publicity and claiming that synthetic chemicals are destroying our way of life." The release listed 10 scientists critical of the claims made in the book; among them was Stephen Safe of Texas A&M, a member of the NAS panel reviewing endocrine disrupters and the most visible of the critics of the hypothesis.

Another scientist listed in the coalition's publicity was Gordon Gribble, a Dartmouth professor of chemistry, who was quoted as follows: "*Our Stolen Future* is like *Jurassic Park.* They each contain a little science and much science fiction. The extrapolations in both books are totally unjustified based on present scientific and medical knowledge. Organochlorines cannot be banned—any more than gravity or photosynthesis—nor should the entire group of anthropogenic chlorinated chemicals be eliminated from our society because of the toxicity of a few." Gribble linked the message in *Our Stolen Future* with the campaign waged by Greenpeace and other environmental organizations against the continued use of organochlorines, a position he disputes in his own book *Chlorine and Health.*

Center for the Study of Environmental Endocrine Effects. According to its public relations materials, the Center for the Study of Environmental Endocrine Effects was established in 1994 "to begin examining the emerging issue of potential effects of man-made or generated 'endocrine disrupters' or 'hormone mimics' on human and wildlife health—principally cancers, reproductive system disorders, fertility, immune system dysfunction, and neurological problems." The center circulates its own scientific reviews, which assess the current state of knowledge and uncertainty regarding the relationship between environmental chemicals and human and animal health. In addition to preparing periodic reports on the issue of environmental endocrine disrupters, the group also views its role as educating the media and developing and maintaining publicly accessible databases on the issue. The center cites its funding sources as both public and private but keeps those sources confidential "to insulate its Science Panel from knowledge concerning sources in order to avoid any possible perception of outside influence."*

American Council on Science and Health. This nonprofit think tank, funded primarily by the petrochemical industry and known as a strong industry voice on science policy, positioned itself to respond critically and sharply to the publication of *Our Stolen Future.* Under the leadership of Elizabeth Whelan, a Ph.D. toxicologist, the council released a point-by-point rebuttal of the book, stating that its thesis is based on speculation and that it emphasizes extreme examples of exposure to estrogenic chemicals, overstates the risks, and seeks to "seduce" the American public with its tale of a maverick scientist pitted against the establishment. "The book is an alarmist tract with a polemical style clearly crafted for its political, not scientific, impact. In a manner reminiscent of its inspiration, *Silent Spring, Our Stolen Future* does not present a balanced picture of synthetic chemicals, which it could easily do by discussing their positive impact on human health and the environment along with their negative impact."[45] On June 12, 1996, Whelan debated Theo Colborn and John Myers at a breakfast organized by Environmental Media Services. Whelan emphasized her opposition to the "precautionary principle." She was quoted in the media advisory as claiming that "Alleged hazards from minute residues of synthetic chemicals have not been conclusively demonstrated. There are man-made and natural chemicals in the environment that have the ability to alter the endocrine system. Yet systematic investigation of this area has only begun and more research is clearly needed."[46] The distinction between natural and synthetic estrogens and their relative degrees of human exposure will eventually play a role in the risk evaluation of endocrine disrupters.

Trade Association Groups. The European Chemical Industry Council, a trade association of about 20 European chemical companies, formed an Endocrine Modulators Steering Group in June 1996. One of its goals is

*The following scientists were listed as members of the center's Science Panel in 1997: William J. Waddell, professor of pharmacology and toxicology, School of Medicine, University of Louisville; William H. Benson, professor of pharmacology and toxicology, School of Pharmacy, University of Mississippi; Keith R. Solomon, professor of environmental biology, University of Guelph; and John A. Thomas, professor of pharmacology, Health Science Center, University of Texas at San Antonio.

to establish a research program with respected scientific experts. The participating companies initially committed about $12 million over a three-year period to research endocrine disrupters. The group planned to investigate the extent to which such observed health effects as declining sperm counts may be due to lifestyle factors, such as diet, rather than chemicals. The group also supported research into the potential effects of phytoestrogens found in such plants as soybeans and nuts; such results could be used by industry publicists to neutralize the postulated effects of endocrine disrupters by linking them to desirable and trusted food products.

In 1995 five chemical trade associations formed the Endocrine Issues Coalition. The group includes the most powerful chemical associations in the world: the Chemical Manufacturers Association with its Chlorine Chemical Council, the Society of the Plastics Industry, the American Crop Protection Association, the American Forest and Paper Association, and the American Petroleum Institute. The coalition set aside $5.6 million for general research that included developing an in vitro yeast system for studying estrogen, androgen, and progesterone receptor binding. Such systems are important as first-tier screening techniques to identify endocrine disrupters (see Chapter 4). The Chemical Manufacturers Association awarded a $1.2 million grant to the Chemical Industry Institute of Toxicology for a three-year research program. Beginning in 1996 the institute allocated $1.5 million of its in-house research budget to endocrine studies.[47] The Chlorine Chemical Council spent over $5 million between 1995 and 1996 on studying the link between breast cancer and DDT, human sperm quality, and wildlife effects of endocrine disrupters. The chemical giant E. I. du Pont de Nemours and Company set aside $2.5 million toward the study of chemical hormone effects.

In January 1999 the Chemical Manufacturers Association announced that under the "Chemical Industry's Health and Environmental Effects Research Initiative," the U.S. chemical industry would spend $1.2 billion on a six-year research program to investigate the health effects of chemicals. More than $500 million was earmarked to test 15,000 chemicals for hormone-mimicking properties. Around the same period, the industry-supported American Council on Science and Health announced that it had convened a "blue ribbon panel" chaired by former U.S. Surgeon General C. Everett Koop to review the safety of phthalates in consumer products.

It was clear that the global chemical industry was building a formidable war chest to combat the theory of endocrine-disrupting chemicals. Public opinion was a key variable in how the controversy would evolve. And one of the newest opportunities for communicating directly with the public was the Internet.

Endocrine Disrupters on the Internet

The Internet became accessible to tens of millions of personal computer users during the same period in the 1990s that endocrine disrupters became a public issue. Advances in computer technology made it possible for home users to network as effectively as well-financed organizations. By 1995 the home page had become every public, nonprofit, and commercial organization's alter ego. Millions of individuals and institutions had registered their Universal Resource Locator (URL) addresses, thus becoming fully authenticated residents of cyberspace. For such controversial environmental issues as endocrine disrupters, the Internet can function as the global version of the Speakers Corner in London's Hyde Park. Anyone with a personal computer connected to an Internet server can speak out on any issue, and his or her voice will be available to the untold millions who are "surfing the Net."

The Internet is a level playing field for any user to advance ideas, disseminate information or misinformation, and advocate causes. Position papers that are posted are not peer reviewed. Informational bulletin boards are not certified for accuracy. There are no gatekeepers for quality assurance of information, as there are in newspapers, journals, or television news. The operative phrase for this ultimate free marketplace of ideas is *caveat lector*—reader beware. Long after media reports fade away and news magazines lose interest in a subject, Internet sites remain. They are available 24 hours a day for months and even years, as long as the creator or manager of the site continues to maintain it. For the Internet, prime time is all time.

For example, in September 1997 I discovered a scientific paper posted on a Web site sponsored by a group identifying itself as the Molecular Toxicology Group, Risk Analysis Section, Health Sciences Research Division of the Oak Ridge National Laboratory. The laboratory is funded by the Department of Energy and managed by the Lockheed Martin Research Corporation in Oak Ridge, Tennessee. All indications were that the site originated from a responsible government agency. The article made some provocative claims and described some new discoveries about endocrine disrupters that I had not seen in the literature: "The Molecular Toxicology Group has discovered that a widely used synthetic food dye, Red No. 3, just like estrogen and DDT, stimulates cultured human breast cells to grow. The amount of this dye in the diet is 1–10 million times higher than the daily total exposure to pesticides or other pollutants like dioxin." It went on to say that Red No. 3 was an estrogen mimic that damages the genetic material of human breast cells. Seven researchers (five of whom were pictured) were associated with the group and the Web site.

I tried to contact the group at the Oak Ridge National Laboratory, only to learn that the agency had been reorganized. The group and section no longer existed, and the Health Sciences Research Division had been renamed the Life Sciences Division. On further inquiry I learned that none of the scientists on the endocrine disrupters Web site was currently employed at the laboratory, and that no one in the newly constituted division would place any credence in the contents of the site. There were no internal documents or published papers that the division personnel could cite on this subject. Within a week of my inquiry the Web site had been removed.

No one yet knows with certainty the influence that Internet sites can have on public attitudes on controversial issues. However, there is a growing expectation that the World Wide Web will be the centerpiece of a new global network of information sources that educate and inform the citizenry. Those whose business it is to shape public ideas and choices have moved aggressively to gain a strategic foothold on the Internet. Given that the issue of endocrine disrupters matured at the same time that the Internet became a new technology for persuasion and education, how was the issue handled? What can we learn about the role of the Internet in scientific controversies? Has the endocrine disrupter episode provided evidence of egregious exaggeration and disinformation?

To pursue these questions I surveyed the Internet in June 1997 for sites pertaining to endocrine-disrupting chemicals, using a series of widely available search engines and applying such key words as "endocrine disrupters," "environmental estrogens," and "hormone-modulating chemicals." A total of 27 individual sites (those defined as having unique URLs) were located: 8 sites of nongovernmental environmental groups, 5 industry or industry-supported sites, 5 governmental sites, 2 academic sites, and 7 miscellaneous sites maintained by private individuals and media groups. Of the 27 sites, 13 took a clear advocacy position.

Controversy over the science of endocrine disrupters and its interpretation is reflected in the political discourse on the World Wide Web.

Home pages from the American Crop Protection Association and the Chlorine Chemistry Council, as well as the Junk Science Home Page, provide useful illustrations. The essential message of the American Crop Protection Association is consonant with that of the moderate wing of the industry. Simply stated, the message is that the environmental endocrine hypothesis is a new scientific controversy that has not yet been resolved. There is little agreement among scientists over its validity. The association's site offers scientific reference articles that emphasize skepticism toward seven postulated health effects: sperm count and sperm quality, prostate cancer, testicular cancer, undescended testes, hypospadia, breast cancer, and endometriosis. According to the site, the environmental endocrine hypothesis is a false alarm—a rush to judgment in which "getting ahead of science can lead to premature, incorrect conclusions."

The Chlorine Chemistry Council was created in 1993 by the Chemical Manufacturers Association. Although the council, often referred to as C^3, has a primary mandate to protect chlorine use, its advocacy extends to chlorinated compounds used in industry and agriculture. Its Web site issued a point-by-point refutation of the claims made under the environmental endocrine hypothesis. The council maintains there is not a single substantiated claim for the thesis, although it focuses exclusively on human health effects and directs most of its commentary to points made in *Our Stolen Future*. From the council's standpoint, diets (such as those rich in dairy products and soya), not industrial chemicals, have exposed people to significantly higher levels of naturally occurring estrogen mimics. The council has reported that the human intake of natural estrogens is several orders of magnitude above that of unintentional artificial estrogenic contaminants.

The council's Web-based counteroffensive against the critics of industrial chemicals extends to America's foremost environmental heroine, Rachel Carson ("Rachel's Folly") and her "intellectual heirs" in the environmental movement. The two great errors in Carson's work as emphasized on the council's home page are her contentions that man's ingenuity (e.g., his ability to create new chemicals) is his worst enemy and that there are no risks in banning chemicals (environmentalists, says the council, consider only the risks of industrial chemical compounds and not the risks created by their absence).

The Junk Science Home Page (www.junkscience.com) was developed by Steve Milloy, listed as executive director of the Advancement of Sound Science Coalition in Washington, a self-proclaimed science watchdog group. Milloy's home page, hosted by a private organization, has as its purpose to poke fun at and discredit a wide selection of scientific claims, including the environmental endocrine hypothesis. The mockery includes a book cover with the title *Our Swollen Future* and a book review with the subtitle "How They Are Insulting Our Intelligence."

The Web sites of environmental organizations vary in their explicit advocacy—some serve as channels of information while others offer the reader options for personal action. The authors of *Our Stolen Future* set up a Web site (www.osf-facts.org) under the supervision of Pete Myers to respond to critics of the book, provide updates on the science of endocrine disrupters, and offer summaries of the key evidence supporting the endocrine disrupter hypothesis. With its interactive component, visitors to the site may submit questions and comments.

Greenpeace International is known for its unique style of creative activism in addressing environmental issues. It has commandeered genetically engineered seeds, blocked fishing vessels and those engaged in illegal ocean dumping, and placed its boats in nuclear test zones. Despite its militant image, Greenpeace chose a moderate tone for its endocrine disrupters Internet site. While clearly sympathetic to the idea that industrial chemicals may be responsible for hormonally induced reproductive problems, the Greenpeace home page on endocrine disrupters is cautious in its claims, recognizes studies that dispute the environmental endocrine thesis, and acknowledges that the evidence is not unambiguous. The following excerpt illustrates the point: "For some time it has been suspected that human sperm counts have been declining, by as much as 50% in the last 50 years [ref.]. Although there seems to be some geographical variation and some studies report no decline in sperm quality, [ref.] recent research has added to the evidence that both the quality and quantity of human sperm are declining in some areas." Greenpeace does not assert that there was a categorical decline in sperm count and quality, nor has the organization accepted uncritically a chemical link to the regional cases of sperm degradation. The site continues: "The key question now is what is causing this [the regional cases of sperm decline]? Could it be the widespread presence of hormonal disrupting chemicals in the environment? Or is there another explanation? Smoking has been blamed by some, as have other lifestyle changes." Greenpeace cites a variety of studies and reports, not all sympathetic to the general thesis. They end with an interrogative, uncharacteristic of an organization often viewed as militantly doctrinaire.

Beginning in 1996, various public and private interest organizations began to stake claims on the Internet, attempting to frame the issue of endocrine disrupters. The industry message was that the hypothesis represented an unproven theory freighted with unanswered questions. Environmental groups situated endocrine disrupters within their existing critical perspectives on synthetic organic chemicals. Succinctly put, their message was that we already have enough information on the probable effects of endocrine disrupters to act, and that existing circumstantial evidence of serious risks justifies precautionary action. The message of U.S. government agencies was that they were applying good science to the issue, that they would meet congressional mandates to screen these chemicals, that they would be undertaking risk assessment initiatives, but that there was nothing to report to consumers at the present. In addition, carefully constructed government sites provided links to industry-sponsored groups such as the Center for the Study of Environmental Endocrine Effects and public interest groups such as Greenpeace and the World Wildlife Fund. Thus, the Internet contained an extensive body of technical, advocacy, and regulatory information on endocrine disrupters, not all of which was dependable or credible. In the hands of a general audience it represented a mélange of claims and counterclaims about tentative health risks. Within five years of the first Wingspread meeting, Colborn, Myers, and others had achieved a notable result: they had put endocrine disrupters on the science and policy agendas of every major industrialized country. To the dismay of some chemical producers and product manufacturers, endocrine disrupter science and advocacy had become the latest environmental cottage industry for research, activism, and policy development.

While industry leaders were developing their strategies in response to the latest environmental threats—with their potential to damage profits, prompt new regulations, and require product substitutions—the media were responding in full force to the new theory.

The Media Response

Media reporting on environmental issues is typically divided into topical areas, including nuclear issues, wildlife protection, toxic spills, and biodiversity. Such neat categorization enables editors to match reporters to the areas about which they are most knowledgeable. A health beat reporter might be assigned to an air quality story, while someone on the natural resource beat might handle a story on the protection of endangered species. The term *environmental endocrine disrupters* was at first a source of confusion to editors, since it defied the ordinary categories. Articles about sperm count decline, reproductive failures in wildlife, the effects of DES, and breast cancer would ordinarily have been treated quite independently. But the introduction of the environmental endocrine hypothesis created a new integrative concept of environmental distress, connecting a wide diversity of events whose mode of action and etiology might be quite distinct. The new integrative concept, while slowly gaining respect within the scientific community, revitalized environmental reporting. It forced journalists and their editors to consider, within a single concept, chemicals that affect animals and humans through any process mediated by hormone modification.

The significance of this integrative concept is that it magnifies the importance of many otherwise disparate, less notable events. In some respects, the term *environmental endocrine disrupter* is for the media analogous to the term *cancer.* Many variant diseases are all categorized under the rubric of cancer because they have in common abnormally formed, unregulated, and invasive cells. Their causes, mechanisms, and outcomes may be vastly different. But having a single concept that unites these variant diseases heightens the public's attention to each individual disease. Gradually, some journalists began to accept *environmental endocrine disrupters* as an organizing principle or lens through which to see the connections among a score of diseases and developmental abnormalities. Pesticides are no longer just toxic or just carcinogens. They are endocrine disrupters. The same is true of industrial chemicals such as phthalates and phenols.

The power that this new organizing concept has acquired in the media and in the public's consciousness is, in many respects, independent of the scientific standing of the hypothesis that incorporates the term *environmental endocrine disrupters.* Because so many outcomes are linked to the term, the mere frequency with which the term is used in the media reifies the concept in the public mind. The reason for this effect is suggested by the following illustration: Five independent disease outcomes weakly correlated to five independent causes do not make a good case for explaining any of the events. But those same five events, each weakly correlated to a *single* cause, paint quite a different picture.

As an organizing principle for science and the media, the environmental endocrine hypothesis began to shape the way in which some environmental stories were presented. Prostate cancer, declining sperm counts, and sexually deviant gulls could be covered in the same article not because there is a single cause, but rather because there is a single idea that connects these events, namely interference with hormone messengers. That idea potentially has the same power as "the mutation and uncontrolled growth of cells" has in our war against cancer.

The media coverage of the environmental endocrine hypothesis began in the early 1990s but expanded substantially after publication of *Our Stolen Future* in 1996. A survey of newspapers and magazines for articles mentioning endocrine disrupters and referring to the environmental endocrine hypothesis reveals the exponential rise of media attention to the issue from the early to the mid-1990s.

A search of the Nexis database was carried out in November 1996 for all news articles mentioning endocrine-disrupting chemicals from January 1, 1990, through November 21, 1996. Records were eliminated from the survey if they were scientific or industry journals, wire service reports, duplicate citations, letters to the editor, government publications announcing congressional hearings, transcripts of congressional testimony, other legislative or executive branch announcements, articles that did not pertain to endocrine disrupters, or transcripts of television and radio broadcasts.*

Our survey of popular print publications indicates that as reporters became more familiar with the environmental endocrine hypothesis, articles appeared with increasing frequency, in more varied types of publications, on non-cancer-related effects of environmental chemicals. The results of the Nexis search yielded 0 articles in 1991, 2 in 1992, 10 in 1993, 67 in 1994, 50 in 1995, and 177 in 1996 (through November).

The environmental endocrine hypothesis broke into the popular media gradually, starting with an article in the Science Times section of the *New York Times* in March 1992[48] and a companion piece by the same author in the *San Diego Union Tribune* in April.[49] Both articles summarized the results of the Wingspread Work Session held the previous year. The pieces emphasized a key point of the hypothesis, namely that environmental chemicals may be causing problems in wildlife and humans beyond cancer.

Of the ten articles on endocrine disrupters published in 1993, the first, which appeared in *USA Today*,[50] discussed the drop in sperm counts, referring to the work of Sharpe and Skakkebaek in *The Lancet*.[51] A flurry of articles appeared in October after scientists had testified at a House sub-

*The exact search string used for the Nexis database search was "endocrine disrupt! or hormone disrupt! or environment! estrogen or estrogen mimic! or xenoestrogen or hormone mimic!" The Nexis library searched was NEWS and the Nexis files searched within NEWS were CURNWS and ARCNWS.

committee hearing on endocrine disrupters. Although the testimony reported in the articles accounted for a potentially wide range of pathological effects, almost all the story leads focused on the breast cancer link to environmental chemicals. For example, a headline in the *Los Angeles Times* on October 22, 1993, proclaimed, "Pesticides May Be Linked to Breast Cancer, Scientists Warn."[52] Reporting from the annual meeting of the American Public Health Association, the *Dallas Morning News* discussed the conflicting evidence on the breast cancer link to environmental chemicals as well as a call by the environmental group Greenpeace for the elimination of compounds thought to be associated with breast cancer.[53]

The number of newspaper articles and magazine stories on endocrine disrupters that appeared in 1994 was more than sixfold greater than that in the previous year. Articles on cancer and noncancer end points were about evenly split over the year. Some articles were prompted by scientific meetings, press conferences, or congressional activity, whereas others were general summary articles produced without any apparent precipitating event. Examples of the latter include an extensive review of the endocrine hypothesis and its critics in the *New York Times*[54] and a three-part series in the *Los Angeles Times* with the provocative title "The Gender War: Are Chemicals Blurring Sexual Identities?"[55] The idea that environmental chemicals might be disrupting reproduction and development was becoming more familiar to journalists. Local and regional reporters used the theme of endocrine disrupters in their reports on wildlife abnormalities. For example, an article in the *Salt Lake Tribune* reported on deformities in birds in Oregon and researchers' efforts to find the cause.[56] The article described evidence of endocrine disrupters and deformities in birds in Green Bay, Wisconsin, suggesting by implication a similar cause.

Two important meetings in early 1994 explain some of the increased media coverage. In January, the NIEHS sponsored a meeting in Washington, D.C., on "Estrogens in the Environment" that was the catalyst for articles in the major U.S. dailies and popular science journals. The articles focused on evidence of adverse reproductive effects in wildlife and humans rather than the cancer link. For example, Janet Raloff's article in *Science News* was entitled "The Gender Benders: Are Environmental 'Hormones' Emasculating Wildlife?"[57] A follow-up two weeks later explored demasculinization in an article entitled "Estrogen's Malevolence: That Feminine Touch. Are Men Suffering from Prenatal or Childhood Exposures to 'Hormonal' Toxicants?"[58] The *Washington Post* ran a story on the NIEHS meeting with a headline that emphasized fertility issues: "Estrogens in the Environment: Are Some Pollutants a Threat to Fertility?"[59] Reports presented at the annual meeting of the American Association for the Advancement of Science in February 1994 resulted in a spate of articles discussing the breast cancer link. These were precipitated by a presentation by Israeli researchers who documented a decline in breast cancer rates in Israel since DDT was banned.[60]

The newspaper and magazine articles that appeared in the aftermath of these two meetings exposed a large segment of the American public to the concept of environmental endocrine disrupters for the first time. They also encouraged public interest groups to develop their own analyses of the current scientific findings. The National Wildlife Federation issued a report, *Fertility on the Brink: The Legacy of the Chemical Age,* which was picked up by papers around the country.[61] Articles in the press noted the federation's contention that low doses of endocrine disrupters could impair reproduction, growth, and immunity, whereas government regulations still focused on high-dose, cancer-causing effects.

The September 1994 airing of the film *Assault on the Male* on the Discovery Channel resulted in several reviews, including one in *People* magazine that warned, "This suspenseful scientific mystery may be the scariest film you'll see on TV this year."[62] The *New York Times* review of the documentary ran in the TV Weekend section of the paper under the headline "Something Is Attacking Male Fetus Sex Organs." The first sentence of the review read: "No, 'Assault on the Male' has nothing to do with the Bobbitt family," referring to the gruesome story of dismemberment by the wife of an allegedly abusive husband.[63] The mainstream American media by and large marginalized *Assault on the Male,* treating it as entertainment and thus demeaning its intended purpose as a serious documentary exposing potential human health risks.

During 1995 the number of published articles on endocrine disrupters declined about 25 percent from the previous year. In January a report in the journal *Developmental Psychology* linking DES treatment to sexual preference in adult offspring caused a flurry of rather sensational articles connecting environmental exposures to gender orientation. For example, the *Arizona Republic* ran a story with the lead "Lesbian Leanings, Estrogen Tied."[64] The issue of sperm decline received a sizable proportion of the media coverage related to endocrine disrupters for several years after *Assault on the Male* was aired. Early in the year the popular press responded to the results of a French study reporting that over the past 20 years the sperm count of Parisian men had declined by 30 percent. The *Atlanta Journal and Constitution* ran an editorial entitled "Falling Sperm Count Sounds Warning,"[65] while the headline in the Cleveland *Plain Dealer* read "Fertility in Males Is Declining."[66]

By 1996 endocrine disrupters had become one of the leading environmental health stories. Decline in sperm numbers and quality was a subject of heavy media interest. Articles appearing in the literary magazines *Esquire* ("Downward Motility")[67] and the *New Yorker* ("Silent Sperm")[68] in January were followed by wide coverage in the daily press. Adding to the media focus on the adverse impact of endocrine disrupters on male fertility were two British reports: one that rats exposed to low levels of the estrogen mimics octylphenol and butyl benzyl phthalate had suffered reduced testicular size and reduced sperm production, and a second one claiming that sperm donors in Edinburgh born after 1970 had experienced a 25 percent drop in sperm count compared with those born before 1959.

The publication of *Our Stolen Future* in March 1996 explains the year's dramatic increase in news stories.[69] Many focused on sperm decline. An article in *U.S. News & World Report* with the headline "Investigating the Next Silent Spring?" reviewed the evidence presented in the book and also cited French and British studies on declining sperm count that were too recent to have been included in the work.[70] *Time* magazine's article about *Our Stolen Future* was titled "What's Wrong with Our Sperm?," and it again focused on the sperm count debate.[71] The *New York Times* ran a review of *Our Stolen Future* along with a companion piece on the sperm controversy: "Sperm Counts: Some Experts See a Fall, Others Poor Data."[72]

The print media showed considerable variation in their reactions to the central thesis of *Our Stolen Future.* The *New York Times* reported the scientific results very cautiously and cited countervailing evidence against the endocrine hypothesis. For example, science journalist Gina Kolata ran a story with the headline "Are U.S. Men Less Fertile? Latest Research Says No," citing two new studies indicating that the sperm counts of Americans were not generally declining and disputing the links between endocrine disrupters and declines in male fertility.[73] Similarly, the *Washington Post* published companion pieces expressing skepticism over *Our Stolen Future* and its implications: "Hormones and Humbug" and "Pollution's Effect on Human Hormones: When Fear Exceeds Evidence."[74]Other less prominent newspapers and magazines basked in hyperbole and drew conclusions well beyond the findings in the book. The *Boston Herald* ran a full-page story titled "Could Men Become Extinct? Pesticides and Plastics May Threaten Male Sex Hormones."[75]

The publication of *Our Stolen Future* brought the terms *environmental estrogens* and *endocrine disrupters* out of the scientific lexicon and into the vocabulary of popular culture. Public interest organizations were quick to incorporate these concepts into their communication strategies against pesticides and toxic substances. Public interest in the sperm controversy, which had been ignited by feature articles in literary and environmental magazines, remained high much longer than might have been the case had the press chosen not to emphasize that aspect of *Our Stolen Future*. Indeed the emphasis on declining sperm counts eclipsed, to some degree, the broader claims made in the book.

Several other events brought print media attention to bear on endocrine disrupters in 1996. A study published in the journal *Science* in June 1996 by a group of researchers at Tulane University found that endocrine-disrupting pesticides acted synergistically to increase their potency.[76] However, when the results of the study were retracted by the authors a year later, there was relatively little press coverage (see Chapter 3). The risk to children was the central emphasis of an event widely reported in the British press in 1996 but largely ignored in the United States. It concerned a scare involving contamination of infant formula with phthalates, which are suspected endocrine disrupters.[77] From July through November 1996, two new pieces of legislation, the Safe Drinking Water Act and the Food Quality Protection Act, drew media attention as they moved through committee hearings. Both bills contained sections relevant to endocrine disrupters, and this was often noted in the press reports.

In summary, since 1992 the media have gradually become more familiar with and increasingly attentive to stories involving endocrine disrupters. At first, signature events such as hearings or new scientific discoveries caught the attention of the press. Although high-profile media events continue to be important in shaping news reporting, the idea of ubiquitous environmental chemicals interfering with reproduction and development has begun to appear more frequently in general articles on the environment. Chemical endocrine disrupters have become part of the accepted discourse in environmental journalism. As a result, many stories that might have been cast as quirky and idiosyncratic, such as the discovery of deformed frogs or sexually aberrant fish, are now placed within the larger framework of the environmental endocrine hypothesis.

Public Perception and Legitimation

The strong and sustained publicity over endocrine disrupters has been perceived by some sectors of the public as a confirmed scientific discovery rather than as a powerful conjecture requiring further testing. Part of this misunderstanding may be traced to the actions of public interest groups, which began using every shred of new evidence to buttress claims that the public's health is at risk from industrial chemicals.

The powerful imagery of the concept of "hormone disturbances" very quickly became incorporated into the public's "worry budget." An illustration of the transition from conjecture to accepted truth is found in a letter to the editor of the *New York Times* criticizing a British pharmaceutical company for exploiting National Breast Cancer Awareness Month. The letter stated that if the company "really wanted to counter the rising incidence of breast cancer, it would stop manufacturing chlorine-based chemicals that have recently been found to disrupt the action of hormones in a woman's body."[78] More than a year before publication of *Our Stolen Future,* environmental groups had already adopted Colborn's view as a working and organizing hypothesis. Writing for the *Amicus Journal,* a publication of the Natural Resources Defense Council, Kenneth Wapner discussed, in matter-of-fact terms, the human health implications of chemical exposures to endocrine disrupters, touching also on the almost *verboten* issue of gender identity: "Theo Colborn's work touches primal fears, casting doubt on our future ability to procreate and our concepts of gender: our reproductive tracts are being altered; synthetic chemicals are being inscribed in our bodies and the bodies of our children and grandchildren."[79]

Popular mass-market magazines exploited the doomsday element of the issue with headlines like "The Incredible Shrinking Man,"[80] "Sex Chemicals,"[81] and "Sex Offenders"[82]—all implying that environmental chemicals were messing up our sex lives. And while these stories might easily be dismissed as so much media hype intended to increase circulation, the journal *Environmental Health Perspectives* began an abstract of one of its papers as follows: "Male reproductive health has deteriorated in many countries during the last few decades."[83]

Other efforts began to build broader constituencies in support of the endocrine model of human disease. The World Resources Institute released a public education report titled "Pesticide and the Immune System." Physicians for Social Responsibility distributed a report, "Generations at Risk," that focused generally on reproductive toxins but incorporated chapters on the effect of endocrine disrupters on future generations. The annual *State of the World,* a report edited by Lester Brown of the Worldwatch Institute, is one of the most widely read and respected publications on global environmental trends. The 1994 volume contained a chapter titled "Assessing Environmental Health Risks" that introduced the subject of endocrine disrupters to subscribers who might generally have been more familiar with such subjects as water quality, oil erosion, population, and biodiversity. With statements like "recent evidence suggests that this load of environmental estrogens could have pervasive influence on human reproduction," the environmental endocrine hypothesis had truly come of age in the international environmental arena.[84]

As we have seen, the issue of chemical effects on hormones is the first major environmental initiative to crest during the breathtaking expansion of the Internet as an instrument for electronic advocacy, communication, and education. Among the most widely cited electronic newsletters is *Rachel's Environmental & Health Weekly,* which is liberally recirculated through electronic mail. This newsletter has followed endocrine disrupters very closely, reporting on new scientific evidence, critiquing media reports debunking the theory, and publishing transcripts of consensus statements from scientific meetings.

A growing number of scientists, in one form or another, have given credence to the endocrine hypothesis by acknowledging that there are both real and potential problems with this class of chemicals. In contrast, industry-oriented magazines have taken the uncertainty over human health effects as license to depict the issue as either unrefined science or fantasy. A clear example of this bifurcation can be found in a 1996 cover story in *Chemical Week,* which ran under the headline "Endocrine Disrupters: Sensationalism or Science?" Despite the tone of this lead-in, the text of the report accepts the scientific grounds for the hypothesis: "While the scale of the threat to human health remains to be established, humans are clearly exposed to the hazard."[85]

An examination of public interest agendas suggests that the environmental endocrine hypothesis has been converted into a framework for action. As is usually the case in symbolic politics, tentativeness in science is a liability for policy change. Many public interest groups, already well organized against persistent organic pollutants, incorporated the endocrine disrupter imagery and research findings to reinforce their existing agendas.

Industry, on the other hand, began emphasizing the importance of "good science," citing uncertainty, the ambiguity of results, confounding variables, and the need for more research. The engagement between science and environmental politics had begun. Which science and which symbols would define the public debate? The next chapter explores the state of scientific skepticism over the emergence of a new paradigm of chemically induced disease and examines the role of contested knowledge in the choice of a policy path.

CHAPTER 3

Uncertainty, Values, and Scientific Responsibility

Change does not come easily in science. Scientists become wedded to certain approaches, methods, or theories, sometimes referred to as paradigms, themata, or frames. These structures define researchable problems—problems that gain legitimacy and set the boundaries of a subfield. Hypotheses or inquiries that do not fit the prevailing paradigm of a subfield have a difficult time finding research funds. And without funding, many proposed experiments that could yield supporting evidence for an unorthodox hypothesis are never performed. This inherent conservatism, while important for maintaining a stable foundation and rigorous standards for scientific progress, may also result in delayed or even forgone discoveries. The physicist Max Planck reported in his autobiography that some of his most distinguished colleagues were reticent about accepting new ideas. He wrote that he had learned a "remarkable fact" from his years of dialogue with other physicists: "A new scientific truth does not triumph by convincing its opponents and making them see the light, but rather because its opponents eventually die, and a new generation grows up that is familiar with it."[1]

Thomas Kuhn described the slow process of fundamental change that takes place in science in his classic book *The Structure of Scientific Revolutions:*

> At the start a new candidate for paradigm may have few supporters, and on occasions supporters' motives may be suspect. Nevertheless, if they are competent, they will improve it, explore its possibilities, and show what it would be like to belong to the community guided by it. And as that goes on, if the paradigm is one destined to win its fight, the number and strength of the persuasive arguments in its favor will increase. More scientists will then be converted, and the exploration of the new paradigm will go on. Gradually the number of experiments, instruments, articles, and books based upon the paradigm will multiply. Still more men, convinced of the new view's fruitfulness, will adopt the new model of practicing normal science, until at last only a few elderly hold-outs remain.[2]

Although not as dramatic an example as the theories Planck and Kuhn have described, a contemporary case in point is the medical hypothesis pertaining to the bacterial etiology of ulcers. Barry Marshall and Robin Warren, two relatively unknown Australian physicians, formulated the hypothesis that gastritis and peptic ulcers were caused by the colonization of *Helicobacter pylori* bacteria in the intestinal tract. They met so much opposition to the idea that Marshall infected himself and became his own experimental subject to make the point.[3] It was over a decade before the Marshall-Warren thesis became generally accepted, after the release of a National Institutes of Health consensus statement on *H. pylori* peptic ulcer disease in January 1994, and the treatment protocol for gastritis and peptic ulcers was radically changed from antacids to antibiotics.

Conservative tendencies in science are also expressed by the reluctance of journal editors and other gatekeepers of the disciplines to entertain grand-scale speculative hypotheses. This reluctance is particularly evident in those disciplines that are heavily grounded in hard-core empiricism. Staying close to the data is a virtue in many experimental traditions, while wandering too far afield is negatively reinforced. This sentiment was expressed by several scientists whose work contributed to the environmental endocrine hypothesis; they indicated that primary journal publications of their research were not the appropriate venue for positing a grand-scale hypothesis about chemicals and disease. It is usually in books, not in peer-reviewed journals, that one finds bold synthetic ideas that cross disciplinary boundaries.

Science, and biomedical science in particular, is also conservative about assertions of causality. Statistical associations and clinical observations are no substitute for double-blind randomized clinical trials. Marshall's published account of his self-induced peptic ulcers and their successful treatment with antibiotics drew considerable attention from the medical community but was not sufficient by itself to change the strongly held beliefs of gastroenterologists that ulcers were a stress-related disease. Marshall and Warren had found *H. pylori* bacteria in the stomachs of almost every patient they diagnosed with peptic ulcers, but that finding still did not establish causality. Skepticism toward their hypothesis prevailed until 1993, when the *New England Journal of Medicine* reported the definitive results of a controlled study showing the success of antibiotics compared with acid blockers and a placebo in the treatment of subjects with peptic ulcers.[4]

The more general and wide-ranging a hypothesis, the more vulnerable it is to skepticism and falsification. The environmental endocrine hypothesis bears many of the characteristics that would lead one to predict an unfavorable or highly skeptical reception by the scientific community. It purports to explain a broad range of human and wildlife abnormalities without a demonstrated mechanism of action that can fully account for many of these effects. There is no critical experiment or definitive set of epidemiological data linking low (ambient) levels of chemicals to a human disease. The hypotheses of cause-effect relations in humans involve highly complex biochemical pathways and multiple chemical agents. The complexity of the biological system reinforces skepticism toward a unified etiological explanation of multiple diseases and developmental abnormalities. Scientists question whether a necessary or sufficient cause is being sought, whether the effects can appear without the putative cause, whether the putative cause can be present without the predicted effect, and whether single chemicals can create the effects alone or whether some combination of chemicals is required.

Before examining the critical responses to the environmental endocrine hypothesis from the published literature and from scientists cited in the media, I introduce a causal framework that can help in sorting out the main areas of scientific skepticism. The framework distinguishes five elements of an explanatory hypothesis that links the exposure of xenobiotic chemicals to disease or reproductive abnormalities: (1) the set of chemical agents, (2) the experimental domain from which the evidence is drawn, (3) the set of postulated adverse effects, (4) the strength of the relationship between some chemical(s) and the observed effect(s), and (5) the nature

Table 2
A Causal Schema for the Environmental Endocrine Hypothesis

Chemicals	*Experimental domain*	*Postulated adverse effects*	*Relationship between chemicals and effects*	*Evidence*
Synthetic organic chemicals	Experimental animals	Cognitive	Causative	Strongly confirming
Phyto-estrogens	Human clinical reports	Behavioral	Associative	Weakly confirming
Metals	Wildlife studies	Reproductive	Correlative	Mixed
	Human epidemiology	Developmental	Circumstantial	Falsifying
	In vitro cell studies	Cancer	Suggestive	
	Biochemical studies	Trans-generational		
		Immunological		

of the evidence in support of or in contradiction to the putative relationship between a chemical and its effect. Permutations among these five elements determine the breadth and scope of the hypothesis and describe the components that make up its evidentiary support, weak or strong. Within the general framework of the theory of chemically induced endocrine effects are sets of nested hypotheses that connect individual chemicals to separate pathologies, not always by the same mechanism (see Table 2).

The Causal Framework

In its most general form, the environmental endocrine hypothesis asserts that certain classes of chemical agents are linked to a variety of adverse effects in both humans and wildlife. The effects in question are associated with interference with the actions of hormone regulators during the early development of or at various stages in the life of the organism. The interference comes about when external agents (xenobiotics) either behave like, interact with, or inhibit the function of the body's natural hormones. The class of agents is identified by several names: *endocrine disrupters,* referring to substances that interfere with the normal functioning of the endocrine system; *xenobiotic estrogens,* referring to chemicals that mimic the role of biological estrogens; and *hormone-modulating pollutants,* referring generally to foreign chemicals that affect hormone levels in an organism.

A set of chemical agents that are foreign to the organism is said to play a role in the appearance of one or more adverse effects in the life of the organism. The exposure may be of short or long duration. Over time, the agents may accumulate in the organism's fatty tissue or blood. Or it may be the metabolites of the chemical in conjunction with other agents that are responsible for the effects. In some cases the effects of exposure to chemical agents may be found in the offspring and not the adult animal, and they may be expressed years after the initial exposure.

The role of science is to evaluate the various hypotheses that claim a relationship between a foreign chemical and a biological effect. Some scientific experiments will yield only weak associative or circumstantial relationships in support of the hypothesis. Other experimental systems may reveal causal relationships, if they exist. Science always favors strong causality over weak causal associations, but in reality the standard of strong causality for the biological effects of chemicals on *humans* is rarely achievable. Policymakers often rely on evidence that is less definitive, such as statistical associations drawn from epidemiological studies.

Science also favors mechanistic explanations for events. In the case of the environmental endocrine hypothesis, a mechanistic explanation involves understanding how the foreign chemical functions in the biological organism to create the pathology. Some scientists believe that a causal link cannot be claimed between a chemical and an effect without an explication of the biochemical mechanism behind the effect. However, most health scientists do not demand that standard of proof to reach the conclusion that a chemical causes a disease.

All evidence in support of a hypothesis linking a chemical to a biological effect must come from some experimental system in which an association (weak or strong, causal or statistical) between the chemical and the effect, if one exists, is discovered. There are preferred disciplinary affinities for certain methods that seek to test whether there is a relationship between a chemical agent and an effect. The experimental systems in use include field studies of wildlife, in vitro cell culture assays, whole-animal studies, human epidemiological analyses, and case-controlled studies.

Those who work with animal models find strong analogies between the way endocrine-disrupting chemicals operate in animal and human systems. A strong linkage found between a chemical exposure and an effect in an animal system might be viewed as providing credible evidence that the effects are likely to be occurring in humans, whereas others, who work exclusively with humans in clinical trials, may be more dubious about such animal-to-human extrapolation. The strongest explanatory systems are those for which there is coherence and consistency across the experimental domains, for example, where there is convergence between wildlife and human epidemiological evidence.

The Agents. The agents referred to by the environmental endocrine hypothesis, endocrine disrupters, are a class of chemicals that either mimic, antagonize, or disrupt the action of endogenous hormones, or alter the synthesis or metabolism of natural hormones in human or animal systems.

The etymology of the word *hormone* can be traced to a Greek word that means to stir up or set in motion. Hormones are a group of chemicals that flow through the bloodstream to all parts of the body and act on cells of target tissues that have receptors specific to the hormone. Thus, hormones activate and regulate our metabolic processes. Scientists sometimes use the metaphor of a lock and key to describe how a particular hormone binds with its receptor and forms a bioactive molecule. Tissues are preprogrammed to initiate events that are signaled or triggered by hormone-receptor molecules. Hormone receptors are classified into two types by their cellular location. The receptors for one group of hormones (polypeptides and catecholamines) are on the surface of cells, whereas the receptors for another group of hormones (steroids) are inside the cells.

The natural hormones of greatest current interest are the steroid hormones (estrogens, androgens, and progesterone) and the thyroid hormones. In principle, however, the environmental endocrine hypothesis includes the entire class of natural hormones in humans or animals—any chemical secreted by any of the endocrine glands: hypothalamus, pituitary, pineal, thyroid, parathyroid, adrenal, pancreas, ovaries, testes—that come into play at any stage of development.

Estrogens represent an important class of hormones in mammalian growth and reproduction. The principal role of estrogens lies in promoting cellular proliferation, in particular the growth of the tissues of the female genital tract and other tissues related to reproduction. Estrogens initiate growth of the breasts and the milk-producing apparatus; they stimulate bone growth and fusion of bones. At least six different natural estrogens are produced in females; three of them are produced in significant quantities. β-Estradiol is the most potent form of estrogen.

Chemical agents identified as *environmental estrogens* are operationally defined in several ways. Scientists use animals to evaluate a chemical's ability to induce cell proliferation in a reproductive organ such as the uterus (uterine growth assay). Female animals injected with estrogens, such as β-estradiol, will develop a thickening layer of cells in their uteruses.

According to Theo Colborn and her coauthors, "The primary model for determining estrogenic activity of a chemical is the stimulation of mitotic activity [cell division] in the tissues of the female genital tract in early ontogeny, during puberty, and in the adult."[5] A sensitive measure of the estrogenic activity of chemicals is the response of the uterus of immature mice: measurable enlargement of the uterus takes place when the mice are fed a diet containing at least 2 parts per billion of β-estradiol. The FDA approved the mouse uterine assay as the official detection method for diethylstilbestrol (DES) in 1963.

In vitro assays are a second means of identifying xenobiotic estrogens. One commonly used assay for screening chemicals for estrogenicity, the E-SCREEN, consists of human breast cancer cells in culture (the strain of estrogen-sensitive cells is identified as MCF-7). In this assay, the estrogenicity of a substance is measured by its capacity in comparison with that of the natural estrogen, β-estradiol, to induce breast cancer cells in a culture medium to proliferate. The determination of estrogenicity by this assay is relatively well characterized: about 50 substances have already been identified that meet the E-SCREEN test. However, there are those who remain unconvinced that it should be used to define estrogenicity. Ashby et al. question the use of the E-SCREEN because they believe that a range of nonestrogenic factors can stimulate those cells to divide and that "not all activities observed in vitro are realized in vivo."[6] According to Ashby, a foreign molecule that can bind to an estrogen receptor in cultured cells raises the possibility, but does not guarantee, that the substance can induce estrogenic responses in the organism.

A set of in vitro assays that relies on estrogen receptor binding consists of variants of a yeast cell with a human hormone receptor attached to the cell's genome. The yeast cell serves as an experimental home for situating the human hormone receptor; a foreign chemical that binds to the receptor in the yeast cell would also be expected to bind to the receptor in the human host. By binding to the hormone receptor, a foreign chemical can induce biological activity in the yeast cell. Scientists can quantify this activity to determine the potency of the foreign chemical for inducing a hormone response in yeast. But is this cell assay a good predictor of what could happen in the living organism? Some scientists, such as Lamb, contend that studies of endocrine effects in cell culture tend to trivialize the complexity of bodily processes. No simple conclusions about endocrine effects can be derived from exposing chemicals to cells because these simplified systems neglect other bodily processes. Lamb emphasizes the complexity of the biological system: "The endocrine system is not just a ligand and a receptor. The body has many levels and modes of control. Receptors are different for different cells. Hormone response elements, which influence cell response, are also diverse. There are controls at the tissue, cellular, and molecular levels. Even receptor levels are not constant in a cell."[7]

Scientists are discovering many biochemical signaling functions of hormones other than receptor binding. The term *cross-talk* among hormonal signals is used to describe the complex web of biochemical interactions that is not reducible to the relatively straightforward mechanism of hormone–receptor–cell–DNA transcription–protein. For example, a chemical that does not bind to an estrogen receptor may interfere with the synthesis of estrogen indirectly by turning off an important gene in the promoter region of the DNA sequence for the hormone.

Currently, there is no universally agreed-upon test for identifying or quantifying the risk posed by chemical endocrine disrupters. Some observers are skeptical over whether a generalized in vitro test or battery of tests can be developed for endocrine disrupters, or even more narrowly for estrogenic chemicals: "Because the effects of an estrogen-like compound can vary depending on dosage, timing, tissue type, organism, and interaction with other hormones and metabolism, a compound may appear innocuous in one test system but produce dramatic estrogenic effects in another."[8] Ashby et al. contend that while "the collective term endocrine disrupter is coming into general use, . . . it has yet to be defined."[9] Table 3, compiled in 1997 by the Illinois Environmental Protection Agency, identifies a preliminary list of 74 chemicals and chemical groups they have classified as known, probable, or suspect endocrine disrupters in humans or animals. A Japanese database from the National Institute of Health Sciences in Tokyo cited 140 chemicals in 1997 for which there was at least one study indicating endocrine-modulating effects in in vitro tests, experimental animals, or wildlife studies. During the same period the World Wildlife Fund, taking a more conservative view, compiled a list of 55 chemicals and chemical groups it classified as endocrine or reproductive system disrupters. In the United States the legal requirement to screen and test endocrine disrupters had prompted a consensus process among stakeholder groups to establish working definitions of endocrine-disrupting chemicals, which resulted in recommendations to the Environmental Protection Agency (EPA) on a battery of tests to identify such chemicals and evaluate their potency. A similar effort had taken place within the OECD (see Chapter 4).

Table 3
Preliminary List of Chemicals Associated with Endocrine System Effects in Animals and Humans or in Vitro

Known	*Probable*	*Suspect*
Atrazine	Alachlor	Aldicarb
Chlordanes	Aldrin	Butyl benzyl phthalate
Chlordecone (Kepone)*	Amitrole (Aminotriazole)	tert-Butylhydroxyanisole†
——— (DDD)	Benomyl	*p*-sec-Butylphenol†
1,1-Dichloro-2,2-bis ethylene (DDE)	Bisphenol-A†	*p*-tert-Butylphenol†
Dichloro-diphenyl trichloroethane (DDT)	Cadmium*	Carbaryl
1,2-Dibromo-3-chloropropane*	2,4-Dichlorophenoxyacetic acid (2,4-D)	Cypermethrin
Dicofol (Kelthane)	Di(2-ethylhexyl)phthalate	2,4-Dichlorophenol†
Dieldrin	Endrin	Dicyclohexyl phthalate
Diethylstilbestrol (DES)*	Heptachlor	Di(2-ethylhexyl)adipate†
Dioxins (2,3,7,8-)	Heptachlor epoxide	Di-*n*-butyl phthalate†
Endosulfans	Hexachlorobenzene	Di-*n*-hexyl phthalate
Furans (2,3,7,8-)	β-Hexachlorocyclohexane	Di-*n*-pentyl phthalate
Lindane	Lead*	Di-*n*-propyl phthalate
Methoxychlor	Maneb	Esfenvalerate
p-Nonylphenol	Mancozeb	Fenvalerate
Polychlorinated biphenyls (PCBs)	Mercury*	Malathion
Toxaphene	Methyl parathion	Methomyl
Tributyl tin	Metiram	Metribuzin
	Mirex	Nitrofen
	p-Octylphenol	Octachlorostyrene
	Parathion	*p*-iso-Pentylphenol†
	Pentachlorphenol	*p*-tert-Pentylphenol†
	Polybrominated biphenyls (PBBs)	Permethrin
	Styrene*,†	Polynuclear aromatic hydrocarbons (PAHs)
	2,4,5-Trichlorophenoxyacetic acid (2,4,5-T)	Ziram
	Trifluralin	
	Vinclozolin	
	Zineb	

Source: Illinois Environmental Protection Agency (1997:3).
*In animals and humans.
†In vitro.

The Experimental Domain. The environmental endocrine hypothesis posits an association between a class of chemicals defined by their functional effects and a variety of health consequences mediated through the human endocrine system. These linkages are postulated and supported through one or more experimental systems, including:

— Epidemiological studies of humans in the general population
— Occupational epidemiological studies of targeted populations
— Animal bioassays
— Wildlife studies
— Clinical observations
— Exposure studies of unique populations
— In vitro cell culture

As an example, various wildlife studies have linked certain chemicals identified as estrogenic (dicofol, dichloro-diphenyl trichloroethane [DDT], and polychlorinated biphenyls [PCBs]) to reproductive abnormalities in specific species. Investigations applying clinical observations also fall within the experimental domain. These methods were used on workers at a chlordecone manufacturing plant to impute estrogenic properties to the pesticide and in studies of women to determine whether residues of DDT metabolites and other pesticides play a role in breast cancer. Yet critics of the endocrine hypothesis argue that occupational studies at high exposures cannot tell us what the effects of low exposures in humans will be, that wildlife studies are not indicative of human effects, and that no human epidemiological studies have discovered a link between low ambient levels of endocrine-disrupting substances and human disease. But *combinations* of experimental domains (e.g., clinical studies at high exposures combined with biochemical assays such as the degree of interaction of an insecticide with uterine estrogen receptors) can have greater probative power than a single domain.[10]

Several of the experimental domains cited have been used to produce toxicological information about endocrine disrupters, particularly their potency and dose-response characteristics. The principles and methods of toxicology grew out of the need to study the human health effects of industrial chemicals, food additives, and drugs. In the early 1900s, Harvey Wiley, the chief of the U.S. Department of Agriculture's Bureau of Chemistry, supervised a group of 12 employee volunteers from his agency who were fed food products with high levels of adulterants, such as the sulfur used in dried fruits. This experiment with human guinea pigs drew considerable publicity and garnered public support for the passage of the first Food and Drug Act of 1906. Although the use of humans in drug testing is essential to evaluate the drugs' efficacy and toxicity, it is generally considered unethical to carry out tests on humans to evaluate industrial chemicals. In recent years, cases have been reported of chemical companies that have tested pesticides on human volunteers, and the EPA is currently considering whether it will accept data from such ethically dubious experiments.[11]

The main issue facing the fledgling field of toxicology during the early decades of the twentieth century was the need to develop methods for measuring human responses to different chemical exposures. Much of the early work was directed at acute toxicity and cancer. Since humans generally were not used as experimental subjects in controlled studies, toxicologists had to devise indirect methods for estimating the dose-response relationships. Animal bioassays, in conjunction with various in vitro assays, have served as the bulwark of the system of chemical regulations in the United States for over 50 years.

The discipline of toxicology emerged with its own set of canons. One of these is that the dose makes the poison. Any substance, natural or synthetic, can be toxic to a biological organism at high enough doses. Conversely, any substance considered highly toxic will be harmless to a biological organism in sufficiently low doses (e.g., a few molecules). A second widely recognized canon in toxicology is that, below the lethal dose of any chemical, higher doses induce greater toxic effects. Dose-response relationships are said to be monotonic—the response rises as the dose increases. If a credible and replicable study shows that a biological organism is not adversely affected by a given chemical at a particular dose, then the conventional wisdom tells us that the organism would not be adversely affected by that chemical at lower doses. Study designs in toxicological research are predicated on such assumptions, at least within certain critical ranges of exposure. Regulations governing the use of industrial chemicals and the clinical applications of drugs are also founded on such assumptions.

However, Frederick vom Saal and Daniel Sheehan have argued that studies of the developmental effects of endocrine disrupters on animals challenge these canonical views in toxicology in predicting low-dose effects of chemicals. Their studies demonstrate that high doses of an estrogenic chemical fed to pregnant mice inhibited the normal development of the prostate in male offspring. In contrast, doses 10,000 times lower administered to pregnant mice resulted in a permanent increase in the prostate size of the male offspring. The authors argued that traditional toxicological reasoning, which assumes monotonic (increasing) dose-response curves, may miss the effects of small doses on the developing organism: "Responses to endocrine disrupters cannot be assumed to be monotonic across a wide dose range, and unique outcomes may occur in response to environmentally relevant doses of endocrine disrupters that may be much lower than doses used in toxicological research."[12]

Toxicologists studying chemicals for acute toxicity often work with dose-response functions that are not well understood in the low dose ranges, look linear or parabolic in the middle ranges, and can taper off like the top of an S in the high ranges. At too high a dose the organism dies. According to vom Saal, on the other hand, dose-response functions for endocrine effects look more like an inverted U than an increasing function. At very low doses the biological response can be high, and at middle to high doses the biological response can approach zero before it rises again. In other words, endocrine effects may shut off when the dose is high enough. Sheehan and vom Saal believe that new thinking is required in endocrine toxicology: "A paradigm shift is in progress. We need to consider some of the important assumptions used in toxicological risk assessments and redesign approaches we use to test chemicals. Such approaches should include use of low doses rather than—or in addition to—the high doses currently tested and should use low-dose-sensitive models.[13]

Extrapolation from the high range of experimentally derived dose-response curves to the low range, as is most commonly done for the estimation of acute toxic effects and carcinogenicity, will fail to give credible results for endocrine effects, according to vom Saal. In recognition of this result, the EPA's advisory committee on endocrine disrupters proposed a screening protocol that begins with a battery of screens to identify whether a chemical is likely or unlikely to possess endocrine-disrupting properties. This initial screen is based not on dose-response characteristics but on endocrine-disrupting mechanisms (see Chapter 4).

The Consequences. More than two dozen effects of xenobiotic endocrine disrupters have been postulated in humans and wildlife, and some have stronger evidentiary support than others. Most of these effects are believed to result from in utero exposure to endocrine disrupters. The best-documented effect of a xenoestrogen on humans is clear cell adenocarcinoma of the vagina or cervix in the daughters of women who were prescribed DES during pregnancy. But unlike most of the chemicals currently under suspicion, DES is a drug. Other postulated effects of endocrine disrupters on humans and wildlife cited in the literature are summarized

Table 4
Postulated Effects of Exposure to Environmental Endocrine Disrupters in Humans and Wildlife

1. Reduction in sperm count/infertility	15. Wasting syndrome
2. Testicular cancer	16. Ovarian cancer
3. Prostate cancer	17. Diminished cognitive function
4. Cryptorchidism (undescended testes in baby boys)	18. Intersex (feminized or demasculinized populations)
5. Abnormally short penises	19. Breast cancer
6. Endometriosis (occurrence of endometrial tissue in the pelvic cavity)	20. Thyroid deficiency
7. Impaired immune system/autoimmune disease	21. Assorted reproductive and developmental abnormalities
8. Abnormal testicular development	22. Goiter
9. Premature breast development	23. Gynecomastia (excessive male breast development)
10. Precocious puberty	24. Decreased testosterone
11. Vitellogenin (egg yolk protein) in male fish	25. Higher embryo mortality
12. Lowered biosynthesis of steroids	26. Hypospadias (abnormal male urinary canal)
13. Eggshell thinning	27. Behavioral abnormalities, such as ADHD
14. Altered sex ratio—diminished males	

in Table 4. With the exception of the clinical uses of DES, U.S. and European agencies that have reviewed the scientific evidence concur with the conclusion of the EPA's 1997 special report that "a causal relationship between exposure to a specific environmental agent and an adverse effect on human health operating via an endocrine disruption mechanism has not been established."[14]

Individual estrogenic chemicals may cause one or more of these effects and do so in one or more species. Alternatively, one or more chemicals may be linked to a single consequence; for example, both PCBs and DDT have been associated with feminizing male species. The predictive power of the hypothesis can be improved and its range expanded in several ways. First, the scope of the causal domain can be increased by the discovery of greater numbers of chemicals that exhibit endocrine-disrupting effects. Second, the linkage between a single chemical and its effects can be strengthened by replicated studies that yield higher statistical power and better case controls. Third, the hypothesis gains power if one can show that a single chemical exhibits one or more characteristic effects across several species. Finally, the hypothesis is further strengthened if chemicals that share one or more properties in common are linked to the same endocrine effects, as when chemicals that show up as estrogenic in in vitro studies are associated with male infertility in occupational studies. For example, the hypothesis that alligators in Florida's Lake Apopka had been estrogenized by pesticides that spilled into the lake from an industrial accident was confirmed by experiments in which alligator eggs painted with a solution of the pesticide in concentrations comparable to the lake exposures showed similar effects on neonates.

Conversely, the environmental endocrine hypothesis is weakened in some of its applications when the evidence does not consistently support its conjectured outcomes. This has been the case with investigations seeking a correlation between organochlorine residues in serum and tissue samples and breast cancer. Human studies have not revealed a consistent pattern. The hypothesis is also weakened when species data are not consistent. For example, if environmental chemicals are responsible for reducing the density or quality of human sperm, it is reasonable to expect analogous effects in wildlife and domesticated animals. If chemicals are adversely affecting human sperm but not affecting that of the animals, some explanation for the differences is warranted. Inconsistency among studies is often an indication that confounding variables are making the experimental conditions different. Thus, two studies testing sperm counts that draw their human volunteers from different groups may reach a conclusion that cultural practices rather than environmental exposures account for the disparity in outcome.

The Linkage Relationship. Discovery of a linkage relationship between two variables is central to the role of science. The methodology and standards for making such a determination are specific to an individual discipline, but in general it will take a cluster of published papers with consistent and replicable results to permit a consensus to be reached that human exposure to a chemical or class of chemicals is associated with a particular effect.

The highest level of confidence expressed by a linkage relationship is *universal (nomological) causality.* This condition is met when the relationship between cause and effect is predictive, universal, and deterministic (whenever the cause is present, so is the effect). In biological systems, however, no two organisms are identical. The same stimulus (e.g., a chemical agent) is inevitably acting on two somewhat different organisms. Whether the genotype of the organism or the time in the person's life cycle during which exposure occurs in fact makes a difference in the cause-effect relationship (the chemical produces an effect in one organism of the species but not in another) is an empirical question. In these circumstances scientists speak of *statistical causality.* Not all women who received DES treatments during pregnancy had young daughters with a rare form of cancer: one out of a thousand daughters were affected. Nevertheless, for those who were, scientists would say that DES caused the cancer.

The minimum condition of statistical causality is that a population sample of the organism, when exposed to the agent under a set of well-defined conditions, will exhibit the same effect to a greater degree (probability) than is observed under similar conditions when the agent is absent. When the relationship between a postulated causal agent and an effect is determined exclusively by statistical methods, without the benefit of a causal mechanism or predictive testing, it is referred to as an *associational relationship.* The power of the statistical association is determined by the size of the sample, the cleverness of the researcher in accounting for confounding variables, and the probability determination that the association could have happened by chance (*p* value). However strong the statistical association is found to be, it cannot claim to be a causal relationship. The term *causal relationship* (deterministic or statistical) applies to variables in a experimental system that can be physically manipulated, under conditions in which predictive and controlled experiments are possible.

For most chemicals, the standard of causality for human effects, especially when levels of exposure are low, is difficult to meet. A principal obstacle to establishing causality is the ethical prohibition against controlled human toxicological experiments with any substance that is not produced specifically for therapeutic purposes. Terms that are used in the literature to signify a linkage relationship weaker than strict causality include *association, suspect cause,* or *correlation.* Linkage relationships between postulated causes and effects also include inferences drawn from analogical reasoning. For example, it may be argued that laboratory tests of animals exposed to xenoestrogens that yield lower numbers of Sertoli cells (testicular cells involved in sperm production) suggest that a similar effect may be occurring in humans. Thess animal studies may explain decreases in human sperm count or sperm quality. Analogy between animals and humans can be more effectively used in building the case for a cause-effect relationship when there is a biological mechanism that is found to be common across species. The mechanism of hormone receptor–mediated responses is among the most promising for establishing the connection between animal and human effects.

Associational inferences derived exclusively from epidemiological studies often draw skepticism from some corners of science for their lack of controls and black-box constructs that offer no mechanism of action.

In science, mechanism rules. Some scientists argue that a strong linkage relationship between cause and effect cannot be developed without a mechanistic explanation. Safe and Ramamoorthy wrote: "Epidemiological studies on wildlife and humans may reveal that a problem exists. Correlation studies that link a chemical to an adverse effect may suggest endocrine disruptors as potential culprits in eliciting adverse effects in humans and animals. But it's at the molecular level that cause and effect must be established if we are to fully understand the mechanism of endocrine disruption."[15] Wolff and Landrigan cite two possible mechanisms by which DDT may increase the risk of breast cancer, only one of which is estrogenicity.[16] Studies that have shown an association between endocrine disrupters and breast cancer are easily discredited by a negative finding and the lack of a testable biochemical pathway of action. Steingraber, on the other hand, contends that understanding of mechanism should not be the guiding principle of our public health policy: "While I am as fascinated as any other biologist at the exquisite nexus of interactions among chemical metabolites, cellular receptors, and DNA, I contend that public health policy has not and should not be driven by the answers to mechanistic questions."[17]

A second limitation in establishing a linkage relationship is that the effects of any single chemical may be cumulative. In some cases the consequences are not manifest for many years and may appear only during certain critical windows (e.g., puberty, pregnancy, or menopause) in the life cycle of the organism. In the DES case, it was decades after the exposure that the effects were first discovered in the adult children of mothers who had been given the drug. When the initiating event is temporally distant from the postulated effect, accurate determination of a causal connection poses unique challenges. As the interval between exposure and effect increases, so does the likelihood that intervening variables may be playing a role. The existence of alternative explanations for the effect also weakens the plausibility of a single temporally distant cause. In the case of breast or prostate cancers, it is possible (and from animal studies plausible) that in utero exposure of the animal to certain chemicals may initiate a cascade of physical effects that will manifest themselves as cancer years later. The amount of the initiating chemical left in the adult animal may not correlate with the risk of cancer. Women who breast-feed their babies release some of the chemicals that have lodged in their bodies, possibly giving a false reading of their level of exposure and risk. Alternatively, the chemical responsible for the initiating cause may have been metabolized. Thus, its persistence in the adult organism cannot be an indicator of exposure. Testing the hypothesis that human breast cancer may be initiated by in utero chemical exposure requires a different type of experimental approach, one that will be costlier than case-controlled studies involving blood and tissue samples taken from breast cancer victims. Such experiments may also take many more years to execute. In addition, some chemicals that accumulate in the body may produce effects that vary depending on where they accumulate and whether there is a threshold effect. If the metabolites of cumulative chemicals are a serious concern, short-term chemical bioassays that exclude metabolites may not reveal important clues.

A third limitation in establishing a linkage between endocrine disrupters and certain health outcomes is the possible additive or synergistic effects of many different chemicals that share some of the same biological activity. If estrogenicity is additive, then the effects of small doses of several chemicals may be the same as that of a large dose of a single chemical. Investigations focusing on one of the many estrogenic compounds lodged in the body may underestimate the total estrogen burden of xenobiotic chemicals.

Synergism is the interaction of two or more agents to produce a combined effect greater than the sum of their individual effects. Chemical synergism may limit the use of field data in studies of wildlife since animals are exposed to so many different chemicals in different formulations, some estrogenic and some antiestrogenic. Testing the endocrine effects of individual chemicals may not disclose the additive or synergistic effects of so-called chemical cocktails. Toxicologists are beginning to take seriously combinatorial toxicities in mixtures. As Altenburger et al. note, "The question remains . . . whether every chemical in any concentration contributes to the overall toxicity of a mixture."[18]

Evidence. In addition to the choice of experimental system and the relationship discovered between the agents and the consequences, the quality, consistency, and breadth of the evidence are all factors that bear on the strength of the hypothesis. Critics of the environmental endocrine hypothesis sometimes point to the selectivity of the evidence advanced for its support. For example, Safe cites the prevalence of plant estrogens (phytoestrogens), which he claims is not factored into the model of endocrine disruption.[19] Skeptics also cite longer life expectancies in industrialized nations as inconsistent with the putative expectation of higher disease rates from exposure to environmental endocrine disrupters. Disease rates, of course, are not the only expected outcome. There are also diminished cognitive function, altered sex ratios, and reduced sperm count—which are not, strictly speaking, diseases.

The relationship between xenoestrogens and breast cancer is one of the subhypotheses advanced within the framework of the environmental endocrine hypothesis that has received the most media attention and, as we have seen, even precipitated congressional action. Early evidence of an association between organochlorines and breast cancer stimulated new lines of inquiry. However, Safe questions the consistency of such evidence, citing data disputing the claim that breast cancer patients have higher serum concentrations of certain xenoestrogens (1,1-dichloro-2,2-bis ethylene [DDE] or PCBs).[20] One of the largest case-controlled studies of plasma organochlorine levels and breast cancer, published in the *New England Journal of Medicine,* does not support an association.[21]

Our Stolen Future draws heavily on animal studies as evidence that there is a strong likelihood of some human effects: "The bare facts are compelling. The animal studies show links between elevated estrogen levels and prostate disease. . . . The DES experience and animal studies also suggest links between hormone-disrupting chemicals and a number of reproductive problems in women, notably, miscarriages, tubal pregnancies, and endometriosis."[22] One of the key issues pertaining to causality is the relevance and validation of the animal model to explain the etiology of human disease. Since the goal of public health–oriented science is to determine the causes of disease or at least its contributing agents, the controversy over the relevance of animal models has played a central role in scientific debates. What happens if the animal models do not stand up to rigorous scientific scrutiny? Do we change the burden of proof? Can we sort out, from among the numerous industrial and dietary agents to which humans are exposed, those that pose a health hazard?

In 1996, Kavlock et al. conducted an extensive review of the literature and concluded there are no "clear relationships" between endocrine effects in humans and exposure to xenobiotics. Only tentative associations were cited. A similar result was reported a year later by a special panel of EPA scientists who affirmed that, except in rare circumstances, a causal relationship between human exposure to endocrine disrupters and adverse health effects has not been established. The report asserted that confirmation of certain forms of the hypothesis is found in the evidence collected from selected wildlife studies. "To date the most credible examples illustrating significant population declines as a result of exposure to endocrine-disrupting chemicals has been reported for alligators in central Florida and some local populations of marine invertebrate species."[23]

Referring to their central thesis on hormone-disrupting chemicals, the authors of *Our Stolen Future* maintain that we must reconsider how we implicate chemicals in the disease process.

> Even if damage is apparent and documented, however, it will never be possible to establish a definitive cause-and-effect connection with contaminants in the environment. Although we know that every mother in the past half century has carried a load of synthetic chemicals and exposed her children in the womb, we do not know what combination of chemicals any individual child was exposed to, or at what levels, or whether he or she was hit during critical periods in their development when relatively low levels might have significant lifelong effects.
>
> ——
>
> We also face the problem of having no genuine control group of unexposed individuals for comparative scientific studies. The contamination is ubiquitous. Everyone is exposed to some degree.[24]

The authors raise a serious obstacle for the scientific method, namely insurmountable barriers to the discovery of the cause of chemically induced disease. Without a generally accepted standard of proof for linking low levels of exposure to a chemical with a human disease, controversy over the standards of evidence for endocrine disrupters played a key role in the critical reviews of *Our Stolen Future*. Commentators on the book were, for the most part, not familiar with the vast number of studies reviewed by its authors. Colborn alone had examined thousands of papers. The critical reviews reveal more about the political response to the new environmental endocrine hypothesis than they do about its scientific merit or plausibility.

Reviews of *Our Stolen Future*

The publication of *Our Stolen Future* was eagerly anticipated by the media as a consequence of the book's prepublication publicity. Environmental Media Services (EMS), with a subcontract to Fenton Associates, handled much of the public relations work. EMS raised money for this purpose from several foundations, not including the World Wildlife Fund or W. Alton Jones. At least two months prior to the book's official release date, references to it appeared in selected publications. Not only had copies of the manuscript been circulating for months prior to the book's release, but it is standard practice for mass-market publishers to distribute unbound prepublication copies of books to journals and magazines. According to Dianne Dumanoski, the chemical industry was preparing a response to pirated copies of the galleys long before the book itself was published.

The reviews of *Our Stolen Future* (appearing in news stories, op-ed essays, and traditional book reviews) were written primarily by scientists, science writers, and staff and editorial writers for news magazines and dailies. Because the book had both a scientific and a policy message, the reviews often reflected the attitudes of their authors toward larger environmental issues. For those predisposed to believe that our chemicalized environment contributes to human illness and ecological degradation, the book was a welcome reminder of the latest chapter in the toxin wars. Alternatively, reviewers inclined to think about environmental issues—and particularly chemically induced diseases—as overplayed used their essays to underscore the antienvironmentalist themes that have become fashionable in the post-Reagan years.

The first reference to *Our Stolen Future* came in a *Los Angeles Times* article in January 1996 by Donella Meadows, a professor of environmental studies at Dartmouth and coauthor of *Limits to Growth,* the influential 1972 report on global resources, population growth, and environmental degradation. Her article, titled "A Chemical Whirlwind on the Horizon," predicted that the book's publication would ignite a controversy surrounding environmental endocrine disrupters. Meadows asserted that the central tenets of the hypothesis were true, and she cautioned readers about the imminent media hype and industry denials, urging them to focus on the widespread deleterious effects of environmental chemicals and to assume that chemicals are "guilty until proven innocent."[25] Other prerelease support came from Jessica Mathews, a senior fellow at the Council on Foreign Relations, who published an op-ed essay in the *Washington Post* on *Our Stolen Future* around the time of its release in early March. Mathews linked endocrine disrupters to a group of environmental contaminants called persistent organic pollutants.[26] This class of chemicals had already been targeted for elimination by several environmental groups under the initiative of a new international convention. Excerpts from the book appeared in the March issue of *Natural History* under the lead "Hormonal Sabotage."[27]

Upon the release of the book, journalists were sent rebuttals prepared by the Chemical Manufacturers Association, the Agricultural Chemicals Association, and the American Council on Science and Health—all industry groups whose member organizations would be adversely affected by restrictions placed on some of the endocrine disrupters. Reviews of *Our*

Stolen Future that summarized the main thesis and highlighted areas of criticism began appearing in the print media a week before the book was released.[28]

From the dozens of reviews of the book, a polarization of opinion emerges regarding its credibility, the scientific standing of the general thesis of endocrine disrupters, and the threat posed by environmental chemicals in general. As science journalist Bette Hileman of *Chemical and Engineering News* noted, "In the popular press, the hypothesis is often characterized as either a confirmed scary reality or chimeric science."[29]

I examined 40 reviews of *Our Stolen Future* published in daily newspapers, weekly news magazines, and science journals between January and September 1996. Of the total, 11 (27.5%) may be considered mixed, 16 (40%) positive, and 13 (32.5%) negative (see Appendix C). Mixed reviews presented a synopsis of the book, including some discussion of the hypothesis along with some points of criticism, without strongly advocating a position. Of the 11 mixed reviews in the sample, 10 were by writers on the staff of the newspapers or news magazines carrying the pieces. The remaining mixed review was an editorial in a science journal. Positive reviews considered the evidence presented in the book convincing and a cause for serious concern. Generally, positive reviews presented little or no criticism of the book and expressed support for the authors' conclusions. Of the 16 positive reviews, 8 were written by scientists or doctors. Science journals or science magazines carried 10 reviews; of those, 8 were written by scientists or doctors. Of the 13 reviewers who identified themselves as scientists or doctors, 8 contributed positive reviews, 4 were negative, and 1 was mixed. Negative reviews communicated strong skepticism or outright rejection of the book's central thesis that chemicals are responsible for human endocrine disorders. Some of the negative reviews went so far as to question the motives and competency of the authors.[30]

To summarize the results presented in Appendix C, mixed reviews of *Our Stolen Future* were most often written by staff writers of daily and weekly news publications, positive reviews tended to be traditional book reviews and op-ed pieces written by scientists and policy advisers, and negative reviews tended to be op-ed pieces written by nonscientist editors and staff writers.

Two of the most influential daily newspapers, the *Washington Post* and the *New York Times*, considered *Our Stolen Future* sufficiently important to publish more than one article, exploring various aspects of the book and its central hypothesis. The *Post* first published a review summarizing the book's thesis, while offering little criticism beyond the observation that readers would be left "with the feeling that the use of toxic chemicals is so pervasive they are doomed whichever way they turn."[31] The *Post* then published a more far-reaching follow-up article on the book, exploring risk perception and the impact on society of scientific hypotheses that predict dire consequences based on inadequate supporting data.[32] One risk researcher quoted in the article concludes that "We're pushing science beyond its capabilities to really settle these questions. And the end result is that the science is better at scaring people than reassuring them."[33] The article examines the behind-the-scenes activities of two advocacy camps: industry groups, seeking to organize a counterattack to the claims presented in *Our Stolen Future,* and the book's supporters, seeking to influence its favorable reception by the public.

The *New York Times* published two articles on March 19, 1996: one reviewing the book and a second focusing on the thesis of falling sperm counts.[34] In May the *Times* ran another article, again exploring the sperm count controversy. This article examined the complexity of the question, Do men today have fewer sperm than their forebears?[35] The *Times* also published a separate and more laudatory review in its Book Review Sunday supplement, in which the reviewer commented that "one of the main strengths of this book is the authors' refusal to let the profound implications of their work propel them into intellectual sloppiness or rhetorical overkill."[36]

Negative reviews emphasized the lack of scientific evidence for human effects, often depicting the claims of the authors as overly alarmist and implausible. For example, the *Los Angeles Times* published a commentary by former Secretary of Health and Human Services Louis Sullivan. Showing no appreciation for the vast wildlife evidence supporting the book's thesis, Sullivan dismissed it as a case built on "innuendo, data extrapolations, and hypothesis," not proven evidence. He disputed the book's claim that sperm counts are declining worldwide, citing evidence that sperm counts may be rising and arguing that the increasing world population is "verification of these findings." Sullivan criticized the wildlife evidence as consisting of isolated incidents resulting from "massive doses of chemicals" and pointed out that fruits and vegetables contain natural endocrine disrupters in higher concentrations than the ambient levels of synthetic chemicals. He argued for setting research priorities based on "real risks" and not the hypothetical risks presented in the book.[37]

Another critical op-ed review published in the *Seattle Times* came from Michelle Malkin of the conservative Competitive Enterprise Institute. She described the work as "a hodgepodge of circumstantial evidence, misleading omissions, and technophobic innuendo."[38]

Gina Kolata used a news story she wrote for the *New York Times* as a book review in disguise. She cast a negative light on the book, concluding that the data it cited did not support its claims. Kolata countered statements by the book's authors with comments from scientists she interviewed, who hammered away at the lack of credibility of the book's central thesis. Arguments by environmental groups that have used the environmental endocrine hypothesis to support their general opposition to synthetic chemicals are described as "fueled more by hyperbole than facts." The review cites experts who believe "there is no factual basis for the book's alarms, several of which have been refuted by studies" and says that "the claims of demonstrable harm, when examined, turn out to be a house of cards."

Bruce Ames of the University of California, a leading critic of the practice of extrapolating from high doses to low doses to evaluate the human health effects of chemicals, is quoted as discounting the antichemical movement—in which he includes the authors of *Our Stolen Future*—as "a political movement . . . based on lousy science." Showcasing Ames, Kolata's review stresses as critically weak points of the hypothesis both the lack of evidence for effects at low doses and the probability that any hormone effects of synthetic chemicals would be dwarfed by the effects of the higher concentrations of plant-derived hormones in the diet (an argument proposed by Stephen Safe). But the most powerful impact of the review is the characterization of Colborn, through her quoted statements, as being unable to defend adequately her own hypothesis. For example, when asked by Kolata to cite the strongest evidence that endocrine disrupters were affecting humans, Colborn reportedly stated that the best evidence came from studies showing that endocrine disrupters were responsible for hyperactivity in children. Then, with stinging irony, Kolata depicts Colborn as having claimed that the evidence for human hyperactivity in children comes from animal studies. What Kolata and others distrust is the use of animal studies to predict human effects. Kolata's rendition of Colborn's responses, juxtaposed with authoritative-sounding criticisms by other experts, leaves the reader seriously questioning the plausibility of the hypothesis and the scientific credibility of its authors.[39]

Gina Kolata herself and the *New York Times* were the subject of an investigative report by Mark Dowie, which sheds light on the quality of the coverage of endocrine disrupters by the *Times*. According to Dowie (and as confirmed by Dianne Dumanoski), in March 1996 Colborn and Dumanoski had presented the thesis of their book to an editorial group at the *Times* that included chief science editor Nicholas Wade:

> When they had completed the briefing, Nicholas Wade, who like Kolata has little patience with presumptive evidence or the precautionary principle, slammed their materials on the table and flew into a rage. "This is not real science. . . . You are creating an environmental scare without evidence. . . . You have no credibility," were phrases recalled by the authors. "Wade railed on for at least two minutes," recalls Dumanoski, an award-winning science writer from the *Boston Globe*. When he was finished with his comments Dumanoski asked him: "Nick, have you read the book?" "No," growled Wade, "I haven't had time." She then asked Philip Boffey, who would be the one to write an editorial if the paper decided to run one, if he had read the book. He hadn't.

Dowie provides examples of selective coverage of endocrine disrupters by Kolata that elevates the status of the critics and ignores responsible supporters of the thesis, while casting *Our Stolen Future* as "hypothesis masked as fact."[40] Yet Kolata was not alone in her impression that endocrine disrupters were a false issue. Ronald Bailey, a television producer, wrote an opinion piece in the *Washington Post* that discredited the environmental endocrine hypothesis and the book's authors. Bailey described the book as "one part pseudo-science, two parts hype, three parts hysteria." He quoted Michigan State University toxicologist John Giesy as dismissing the book and its primary author: "Frankly, Colborn doesn't know very much. She reads the entire literature and picks and chooses things that support her preconceived views." Giesy disputed the evidentiary foundation of the book: "There is no evidence of any estrogens causing problems in wildlife."[41] One might assume from Bailey's quotations of Giesy that the toxicologist does not think too much of the hypothesis. However, a different picture emerges in an article on endocrine disrupters that appeared in *Chemical and Engineering News*, in which Giesy is reported to have said, "When you scrutinize the data it doesn't seem like there are widespread effects in human populations. The compounds cause effects at very low dose and there is reason to be concerned, because in wildlife we have some effects that can be attributed to contaminants."[42]

Positive reviews emphasized the plausibility of the hypothesis without highlighting criticisms or uncertainties that had begun to appear in the scientific literature. The book was used by some reviewers to launch a broadside against the chemical industry for its abuse of the environment.[43] To these critics, the hypothesis was just one more indication of how bad things continue to be vis-à-vis chemical pollution and the declining health of living things.

In a more balanced review, the editors of *Environmental Health Perspectives*—the journal of the National Institute of Environmental Health Sciences (NIEHS), which has published a number of research articles on endocrine disrupters—pointed out that the book was not a dispassionate treatise: "Readers . . . should be mindful . . . and recognize that *Our Stolen Future* is neither balanced nor an objective presentation of the scientific evidence on . . . whether exposure to endocrine-disrupting environmental chemicals is causing significant increases in human disease." Nevertheless, the editors commended the authors for bringing together and updating "considerable toxicological evidence that environmental substances that mimic or block actions of hormones are producing many of the widespread adverse effects on reproductive capacity and development in wildlife." They argued for "science-based risk research" in a number of "critical areas" related to endocrine disrupters to provide regulators with "key elements in making decisions in the face of uncertainty."[44]

Anthony Cortese, the director of a nonprofit environmental education organization, former dean of environmental programs at Tufts University, and past commissioner of the Massachusetts Department of Environmental Protection, stated in his review that the evidence from laboratory and wildlife animal studies and the DES data presented in *Our Stolen Future* were cause for concern. The focus of Cortese's review was the impact of the hypothesis on public policy. He argued that the hypothesis "creates some fundamental scientific and policy challenges" and considered the authors' paramount message to be "get beyond the toxicity equals cancer paradigm." He noted that endocrine disrupters are not classic poisons or carcinogens and called the authors' public policy and research recommendations sound. Cortese warned against waiting for overwhelming evidence and applying the "OK until proven guilty" paradigm for chemical safety.[45]

Writing favorably about the book in *BioScience,* Maurice Zeeman of the National Institutes of Health stated that the authors' claim "that hundreds of billions of pounds of industrial chemicals . . . are made in the United States every year . . . is off by at least an order of magnitude—too low!" Zeeman claimed that these numbers do not even include pesticides, pharmaceuticals, and human and animal food additives.[46]

The response of the media was in some ways predictable in its general outlines. Most reviews by staff writers for the major daily and weekly news publications took a middle-of-the-road position, balancing the conclusions of *Our Stolen Future* with statements from its critics, thus remaining true to the traditional role of journalists. More partisan perspectives came in the form of opinion articles published by those individuals with well-established positions on the safety or hazards of environmental chemicals. Many of these reviewers, whether supportive or critical, focused on those aspects of the central hypothesis that reinforced their own perspectives on synthetic organic pollutants. As might be expected, there were no "born-again" environmentalists among the authors of these 40 reviews.

Vice President Albert Gore wrote in his foreword to *Our Stolen Future* that the book "takes up where [Rachel] Carson left off."[47] He was not the first to draw comparisons between Carson and Colborn and between their respective books and life experiences. Yet the scientific community treated Carson quite differently than Colborn and her associates. In part this may be attributed to their different backgrounds and their distinctive professional associations within the community.

Rachel Carson received her master's degree in biology and taught for several years at the Johns Hopkins University and the University of Maryland before taking a civil service position with the United States Bureau of Fisheries. But her passion was for writing about the natural world, which she did first in literary magazines and then in a highly acclaimed book titled *The Sea Around Us,* which was on the best-seller list for 86 weeks, was chosen for the Book-of-the-Month Club, and was condensed in *Reader's Digest.* She received her highest recognition in literary circles for her essays in the *New Yorker,* which serialized chapters from her books on the sea. Thus Carson was a nature essayist before she undertook the writing of *Silent Spring,* in which she put forth a scientific argument, cast within an ethical framework, against the use of many forms of synthetic pesticides.

In contrast, Colborn received her doctorate and never taught; however, like Carson, she began her environmental career as a scientist working for a government agency. From the outset, Colborn developed close ties with members of the scientific community, creating ever-widening circles of contact. She began publishing scientific papers and editing symposia proceedings years before considering a popular book on chemicals. She was quickly accepted into the community of scientists and respected for her integrative ideas and her ability to cross-fertilize scientists from diverse disciplines on issues of wildlife decline and human health effects.

These variations in the career trajectories of their authors may explain, in part, the different critical responses to the two popular treatises on chemicals and health. It must also be recognized that it was Carson who dispelled the myth of chemical modernization, namely that synthetic chemicals were representative of industrial progress, that science could do no wrong, and that nonexperts had nothing to contribute to our understanding of chemical dangers.

In anticipation of the release of *Silent Spring,* one major chemical corporation, by threatening a lawsuit, tried to persuade Houghton Mifflin to reconsider its decision to publish the book, citing gross inaccuracies pertaining to the company's pesticides chlordane and heptachlor. In addition to the anticipated attacks on the book by chemical and agricultural companies, *Silent Spring* drew vitriolic criticisms from academic scientists, who cited Carson's lack of standing in the scientific community. The review in *Chemical and Engineering News* was emblematic of the harsher criticism: "Her ignorance or bias on some of the considerations throws doubt on her competence to judge policy."[48] However, Carson also received some balanced and sympathetic responses from scientists, including one who reviewed her work in *Scientific American.* While questioning some of her claims, the reviewer did not accept uncritically the safety of pesticides. According to Carson biographer Pat Hynes,

> The attacks on *Silent Spring* were very often attacks on Carson. They associated her with communists and "nature fanatics." They attributed idiosyncratic tendencies to her and made misogynist remarks about her person, personality, and lifestyle. She was portrayed as an eccentric spinster who liked birds better than children. They charged that she was unqualified to author the book. Some said she was not a scientist; others said that even if she were, her writing was too emotional to be objective. Her emotions blinded her to the progressive nature of chemical pesticides. In blocking progress, she was endangering the planet.[49]

Even as the lead author of *Our Stolen Future* and its scientific heroine, Colborn was not the target of the type of overt criticism Carson had weathered. Industry spokespersons and apologists were far more restrained in their reactions to Colborn and her book. The scientific reviews were much less divisive and avoided the kind of harsh, ad hominem attacks to which Carson had been subjected. Some scientists criticized the book for being too folksy, for lacking rigor, for its tendency to mix anecdotal evidence with rigorous experimental results, and for failing to consider negative studies. The *Scientific American* review, written by Michael Kamrin and titled "The Mismeasure of Risk," faulted the authors for presenting "a very selective segment of the data" and for the adversarial position they had taken, the latter being a violation of science's positivist creed: "The book is not scientific in the most fundamental sense, because it aims to convince readers about what ought to be rather than to explain what is or what is likely to be."[50] Yet throughout the book's reception, Colborn was buffered by a circle of scientific supporters—a number of them Wingspread alumni who had had an opportunity to review sections of the book in draft form—who were willing to stand behind the general thesis that environmental endocrine disrupters could be the cause of human reproductive and developmental abnormalities and to argue that the thesis should be thoroughly investigated.

Whatever ambivalence there was in the scientific reception of *Our Stolen Future* is best expressed by a review in the journal *Science*. The first sentence of the review states "this book was not written for scientists"; by the last paragraph the authors write that "the potential threat of endocrine disrupters is a critical issue for our time." The power of the hypothesis had been acknowledged. It had taken some eight years to bring it to the attention of the public and the broad scientific community. Before Wingspread I in 1991 the term *endocrine disrupter* was simply not in our lexicon. Now, with the concept of endocrine disrupters appearing on the radar screens of scientists and the general public alike, would science and government address its concerns? The *Science* review posed the challenge: "Scientists must decide whether we should contribute to the ongoing debate solely by providing data or whether we should also recognize and accept the responsibility to participate in the equally important component of policymaking that involves rendering of value judgments."[51] Scientists responded to this challenge in different ways.

The Social Responsibility of Science

Science is far from a homogenous command-and-control system. It thrives on controversy, skepticism, and open debate. But science also has ways of reaching consensus, particularly among those who have attained the position of gatekeepers or leaders in their respective disciplines.

Consensus translates into the codification of methods and principles into canonical texts. Published journal articles on a scientific hypothesis do not imply that the results cited are generally or even widely accepted, that the outcome is generalizable, or that the experiments from which the results were obtained are replicable. Failure of a paper to gain entry into the major journals of its discipline usually signifies that the science described in the paper is deficient or that the subject matter is not important enough to be published in that journal, according to its reviewers, editors, or both. Scientists who have reached a certain level of sophistication understand that rejection of a paper by one journal does not imply that it will not be favorably received by another of perhaps lesser or even commensurate stature.

When controversy persists in an area of science, the competition to advance the acceptable theory continues until the evidence builds in support of one leading explanation. If science is to be rational, it must be endowed with patience. It is not the function of science to select truth from among equally untenable hypotheses.

It is useful to distinguish two types of controversies in science: scientific disorders of the first and second kind.[52] Disorders of the first kind pertain to conflicts and anomalies within science's internal development. Examples include a theory confronted by anomalous data or two explanations competing for hegemony in a divided discipline. Disorders of the first kind originate from within science and often bring critical attention to foundational issues. At one time there were disputes over whether space was a plenum filled with a substance called ether and whether AIDS was an autoimmune disease—hypotheses that have since been discredited.

Disorders of the second kind arise when science is outer-directed, that is, when its theory and methods are used to address problems arising out of a social context. The application of science to evaluate risks falls into this category. Science may be sought out by policymakers to resolve a concern about human health or ecological effects. In such cases the policy sector may not be in a position to wait until the uncertainty has been narrowed before it makes its decisions. Even a non-decision is a decision, since it may mean the continued release of a toxic substance or the exacerbation of a demonstrated effect such as atmospheric ozone depletion.

The term *second-order* or *postnormal science* was introduced by Funtowicz and Ravetz to describe the application of scientific approaches in the face of high risk and high uncertainty.[53] The central features of postnormal science are that the facts are uncertain, the values are in dispute, the stakes are high, and the decisions are urgent. Under such conditions it is "soft" scientific information that serves as inputs to the "hard" policy decisions regarding many important environmental issues.

Scientists face a different set of ethical problems between first-order and second-order science. In this section I explore the notion of scientific responsibility for scientific disorders of the second kind, with special attention given to issues raised by endocrine disrupters. Here are the central questions:

— How do scientists interpret their ethical responsibility to society when faced with an environmental hypothesis of high stakes and high uncertainty?
— What is the moral criterion adhered to by scientists who become advocates of social change while awaiting definitive evidence on the status of an environmental health hypothesis?
— How should scientists communicate with the media when faced with an uncertain hypothesis linking an environmental cause to a health effect?

The ethics of engagement pertaining to scientific controversies of the second kind are not codified, nor are they part of a set of institutional norms. In contrast, the ethics pertaining to science as it develops internally (first-order science)—for example, rules governing the use of human subjects and animal experimentation—have become a routine part of scientific practice. An emerging ethical issue that bridges first- and second-order science is conflict of interest. Scientists with financial interests in their research are increasingly being called upon to disclose those interests in grant applications, book reviews and editorials, research papers, and public hearings.

The sense of personal responsibility held by scientists toward environmental health hypotheses depends on several factors: how they view their role in science; how they view the role of values in science; and how they view the role of science in society:

— *Traditionalist versus nontraditionalist.* The traditionalist scientist sees his or her role exclusively as a seeker and communicator of knowledge. For this individual, it is the role of other sectors of society to decide how that knowledge should be applied. Traditionalists avoid drawing any moral or political implications from science that may inform public policy. They also avoid straying too far afield from their areas of expertise.
— *Positivist versus nonpositivist.* The positivist believes that science is a value-free pursuit, that scientists operate within a set of ethical standards to ensure that the integrity of research is protected, and that the normative structure of science can serve as a model for social decisions.

In this tradition, controversies in science must be resolved from within. Moreover, public controversies that rely on scientific information must draw their resolution from the best scientific authorities. For the positivist, technical controversies can be segmented into pure scientific and pure value issues. The former should be resolved by experts before society, through its democratic processes, addresses the latter.

— *Public versus private.* The public scientist is someone who is comfortable communicating through the mass media. Some scientists have a principled distaste for speaking to the press. They offer several reasons for this aversion, including the following: the media oversimplifies or distorts scientific messages; quotes from scientists are taken out of context to serve the opportunistic goals of the media and compromise the serious pursuit of truth. Another reason some scientists give for avoiding public discourse is that the popularizers in science are not taken as seriously as the purists. In becoming public scientists, they run the risk of losing credibility among their colleagues.

— *Government-, industry-, or nonprofit-supported.* The organizational affiliation of scientists is also a key factor in how they view their social responsibility when faced with a publicly volatile issue. Government and industry scientists rarely become involved in public controversies because it is difficult for society and those to whom they report to disassociate their personal views from those of their organizations. This is also true for scientists who are employed by public interest groups, except that in this case they are hired because their personal beliefs and those of the organizations that hire them are expected to be compatible. Universities are nonprofit institutions that are not expected to have a policy point of view. It is generally understood that academic faculty speak only for themselves. For this reason, American universities are seen as reservoirs of independent minds, whose positions on public issues are not shaped in any direct way by the interests and values of their institutions. The sources of funding of academic scientists may be a better predictor of their perspectives in areas of contesting knowledge than the "points of view" of their institutions.

The Ethics of High-Stakes Hypotheses

Scientific conjectures can create conflict, fear, and anxiety in society, whether the conjecture is eventually found to be true or false. I am particularly interested in early-stage hypotheses—call them nascent or immature, half-baked, evolving, or speculative. If science were only internally directed, then Karl Popper's view—let a thousand conjectures bloom before we begin the process of refutation—would have no special moral implications. But some hypotheses can foment, disturb, create sudden or radical behavioral changes, and result in economic dislocations.

We have seen entire communities abandoned on the conjecture that toxic waste threatened the health of residents. Widespread fears have circulated about the effects of mercury amalgams used in dentistry, about natural radon in our homes, and about the effects of high-intensity electromagnetic fields in our neighborhoods. Each of these issues is associated with one or more hypotheses linking an environmental agent with a health effect. There are ethical consequences to taking a policy position on these hypotheses. We may react prematurely at great cost and personal hardship. We may wait for more information and unwittingly permit the continuation of harm. Or we may eliminate one environmental threat while inadvertently introducing another.

Among the ethical problems that arise in connection with the introduction of nascent hypotheses are release of uninterpreted data, conjecture sensitivity, prepublication release of research, and conjectural risk.

Release of Uninterpreted Data. Ethical dilemmas associated with the release of uninterpreted data have been most visible with respect to the data from cancer registries. Several states, including Connecticut and Massachusetts, collect cancer incidence data that are disaggregated at multiple geographical levels. Inevitably, elevated tumor rates for some sites show up in communities. How should these data be handled? Should communities be alerted to the elevated sites despite the lack of any reasonable explanation for the higher incidence of a specific cancer? What is the responsibility of public health scientists in the face of information about elevated disease rates, without knowing anything about the microdemographics of the community (e.g., smoking rate, alcoholism, residential migrations), particularly when the numbers may be very small? For example, in Massachusetts 21 of 351 municipalities had breast cancer incidence rates among females 20 percent higher than the state average over a 10-year period. The release of this uninterpreted information, and the additional knowledge that 9 of the 21 sites were among the 15 towns of Cape Cod, prompted strong community responses, leading to the establishment of a state-funded study on the causes of breast cancer. The publication of cancer registry data in Massachusetts was thus a catalyst for the organization of breast cancer activists, who lobbied for investigations into the environmental causes.

The legal requirement that pure environmental risk data be published obviates the ethical dilemma associated with the question of whether or not they should be released, but it also pushes the ethical problem to another level. What is the responsibility of public health officials and scientists in the wake of reports of elevated rates of disease? Where is the burden of responsibility for finding the explanation for the elevated disease rates? If an elevated rate of an infectious disease, such as a coliform outbreak, were found, would the report of the outbreak not be followed by a search for its cause? The public health community is poised to move quickly and strategically on reports of infectious disease outbreaks, but it responds selectively and by virtue of political pressure to reports of elevated cancer rates. We have more difficulty as a society understanding our obligations toward elevated rates of cancer than we do toward elevated rates of infection.

What about the release of disaggregated data? Should cancer registry data be available block by block? Toward what end? According to what standards do we release data? Is there an obligation on the part of the public health community to anticipate how communities might react and use that knowledge in determining how the release of information should take place?

Ethical concerns about the release of data are also evident in the field of drug testing. Consider the case of a publicly funded clinical trial in which a placebo and two drugs are being tested for effectiveness in treating HIV infection. Let us suppose that at some point in the middle of a clinical trial an AIDS patient initiates a legal action to obtain access to the raw data, recognizing that it is incomplete, in order to make a last, desperate effort to survive the disease. Should the early-stage unpublished data from this publicly funded research be accessible to people outside the research team? Who should have control of the data in publicly funded research? In high-energy physics, teams of scientists generate raw data that are not released to other physicists unless they are first published by the team that generated them. The convention in this field has long been that access to raw data is limited to the members of a high-energy collaboration (which could consist of hundreds of scientists), on the assumption that only those who generate particle physics data can understand how to interpret them. But in the field of cancer epidemiology communities are given access to tumor registry data, and their interpretation of them becomes part of the public dialogue.

In her book *The Apocalyptics,* which describes the misdemeanors of regulatory science, Edith Efron raises the problem of releasing data on carcinogens before there has been proper peer review. Efron cites two National Cancer Institute scientists who describe the sensitive ethical position in which researchers with socially important data find themselves: "A delicate ethical dilemma confronts scientists in selecting the proper timing and extent for the release of carcinogenesis test results. Early findings may not be confirmed and may cause technological and economic problems and unnecessary anxiety. On the other hand, delaying public notification of highly suspicious findings until a final detailed report is published may delay preventive actions that could protect exposed populations from unnecessary risk."[54] Even the careful publication of chemical toxicity data does not guarantee that the results can be replicated, or that they will be consistent with a more extensive and thorough analysis of the chemical in which greater emphasis is placed on eliminating confounding variables.

Conjecture Sensitivity. There are those scientific hypotheses, theories, or investigations that evoke intense degrees of discomfort among certain groups. This was the case with Charles Darwin's theory of natural selection, Arthur Jensen's thesis on the genetic basis of intelligence, E. O. Wilson's theory of sociobiology, and, more recently, conjectures about genes and criminal behavior. The reasons for this discomfort are varied. Certain conjectures may conflict with deeply held religious or social convictions, or they may reinforce social divisions or racial prejudice.

Many scientists take a libertarian view with respect to pure scientific inquiry, according to which the right to establish conjectures and to pursue evidence in their support shall not be constrained by government or popular opinion. For example, in the field of environmental health, some behavioral hypotheses may reinforce a "blame the victim" approach that undermines prevention strategies. Since the public health agenda has begun to include such issues as violence prevention as a means of health promotion, some public health advocates question the morality of disseminating speculative theories that could undermine social programs and stigmatize certain minorities.

A National Institutes of Health–sponsored conference on "The Meaning and Significance of Research on Genetics and Criminal Behavior" provides an interesting case of conjecture sensitivity. The objective of the conference was to explore hypotheses suggesting a genetic cause for violent crime. Following that logic, prevention would involve either fetal screening and abortion or screening of newborns in conjunction with some intervention such as monitoring of behavior. Thus, the social implication of the hypothesis might be the suspension of civil liberties for those with the genetic marker, even when they have not committed a felonious act. Some scientists and bioethicists contend that research on the relationship of genes to behavior has no place in our society because it assumes the individual is predetermined to a life of crime.[55]

Should scientists exercise restraint in pursuing highly speculative, socially divisive hypotheses, or is this notion of social responsibility anathema to the spirit of free inquiry? A similar issue arose in the late 1960s when a research team at the Harvard Medical School was investigating a hypothesis that men born with an extra Y chromosome (known as XYY men) had a predisposition toward criminal behavior. The research was abandoned in the wake of protests from colleagues who criticized the methodology of the study and opposed the ideology underlying the hypothesis. These debates reminded me of the German intellectuals who had backed theories of racial superiority that were eventually used by fascists to lend a scientific veneer to their social policies of ethnic cleansing.

Prepublication Release of Research. Releasing new scientific results to the media prior to publication and independent peer review has been the subject of controversy in the biomedical sciences. Two typical motivations behind the release of prepublication scientific results to the press are to gain commercial value or achieve political advantage from the discovery. In 1980 the Cambridge, Massachusetts–based biotechnology company Biogen held a press conference announcing its successful microbial production of interferon, an antiviral agent, by the application of recombinant DNA techniques. Biogen's news release was issued before the results were published in the scientific literature. An editorial in the *New England Journal of Medicine,* titled "Gene Cloning by Press Conference," cited the break with tradition and the disregard of professional responsibility implied by prepublication communication of scientific results to the media. "In place of published data, open to all for examination and critical review, we now get scientific information by press conference."[56] Some scientific journals will not accept submissions of original research released to the media before they have been peer-reviewed, accepted for publication, and published.

A second example of unpublished scientific results released exclusively to the media creating a stir was the Alar controversy. The Natural Resources Defense Council (NRDC) released a study to the media on the cancer risks to young children from exposure to 23 types of pesticide residues found in food. The study had been conducted by NRDC staff scientists with the help of scientific advisers and a select group of peer reviewers. Working with the public relations firm Fenton Associates, the NRDC offered the television news program *60 Minutes* an exclusive on the report in exchange for a feature story. Scientists received their first glimpse of the cancer risks of Alar residues in food in journalistic accounts published after the story had aired on prime-time television. The rest is history. The producers of *60 Minutes* had selected Alar from among the 23 pesticides as the "hook" to frame the story. Public reaction was swift and decisive. Alar became stigmatized and within several months lost its commercial markets. Some cried foul. The EPA's scientific advisory committee, after all, had not found sufficient evidence to convict Alar of toxic trespass in children's diets. The NRDC, in turn, cried foul when it learned that the majority of the EPA's advisory panel had consultant relationships with Alar's manufacturer, Union Carbide. Mainstream academic journals used the Alar incident to emphasize the importance of peer review and to remind the media of its responsibility in reporting on the "gray" literature—the term used to refer to nontraditional, non-peer-reviewed publications.

Conjectural Risk. The Alar episode leads me to the final type of ethical problem, conjectural environmental risks. What forms of social responsibility are expressed by scientists when they are confronted by a conjecture about hazards to humans or the natural environment? How does the culture of science contribute to the scientists' sense of social responsibility with respect to conjectures of high stakes and high uncertainty? Examples in this category are manifold and include deterioration of the ozone layer, global warming, electromagnetic fields, radon, and, most recently, chemical endocrine disrupters.

There is a wide variation among scientists in how they frame their social responsibility with regard to hypotheses about environmental risk. Some scientists view their role as limited to producing research. They avoid making any policy recommendations, such as proclaiming what is acceptable risk or when government should take action. For this group of scientists the function of hypothesis formation is strictly to carry on science. They see the pursuit of science as an intrinsic good that does not depend on whether it informs or influences public policy.

Other scientists see hypothesis formation in an instrumental sense, namely as a means of protecting society from the untoward effects of technology. They will bring to the public's attention purely speculative risks that call into question the safety of a new product or process. These hypothetical scenarios may have little or no empirical evidence backing them, but they are at least possible and at most plausible events. Scientists who use hypotheses in this way view their role as providing a protective buffer between the purveyors of technology and the rest of society by offering early warnings of potentially hazardous outcomes. Their sense of professional responsibility is linked to supporting rigorous assessment prior to the manufacture and release of a product or the introduction of a new technology.

In 1971 a highly influential hazard scenario involving the transfer of animal tumor viruses to bacteria touched off the recombinant DNA controversy. The conjecture raised the possibility that the transfer of monkey tumor virus DNA into bacteria could spread these viruses into the human population.[57] When the risk scenario was conveyed to the popular press, scientists were divided on the ethics of arousing the public exclusively on the basis of theoretical risks. Scenarios such as this helped draw attention to the potential risks of gene splicing technology and touched off what has been called "the gene wars."[58]

Scientific risk conjectures are not all purely speculative. Among them we can find the entire spectrum of evidentiary support. The central questions are: When is the evidence sufficient for scientists to initiate a call to action? When is such a call to action seen as inflammatory science or "junk science"? And when is it seen as "prudent science"?

The risks of natural radon were calculated through the use of predictive modeling based on exposure data from epidemiological studies of more than 40,000 miners. The cancer risks of Alar were calculated by extrapolating human risks from animal data. Although the EPA took a very weak position on Alar, the agency made radon the basis for a national campaign. In the 1980s the EPA estimated that 10,000 additional cases of lung cancer annually resulted from home exposures to radon gas; excess cancers due to Alar were estimated by the NRDC at about 5,000 per year. By 1996 Americans had spent over $400 million testing for and mitigating the perceived effects of radon. New studies of residential exposure raised some doubts about the initial estimate of radon risks,[59] whereas a reanalysis of the uranium miner data by the National Academy of Sciences estimated that indoor radon contributed to between 15,400 and 21,800 of the 157,400 lung cancer deaths reported in the United States in 1995.[60]

Do high uncertainty, high-stakes hypotheses require a bimodal system of ethics, namely, a system with one set of principles applied to science and another set to political action? Should we allow or require science and scientific epistemology to drive the political and regulatory process? This view seems to be gaining popularity among conservative law-makers, who argue that we should not regulate until there is sufficient scientific evidence of a risk. In their view, the threshold for regulation is based on the threshold for scientific publication of a cause-effect relationship. Can we afford to allow the burden of proof as determined by scientific epistemology to establish when we intervene, when we regulate, and when we mitigate?

In some cases, such as lead in gasoline and occupational exposure to asbestos, the problems were worse than initially thought. In other cases, such as electromagnetic fields and the laboratory hazards of recombinant DNA research, the risks were less serious than some originally postulated. What is the appropriate level of proof before the responsibility of action is taken? Must law and regulation be based exclusively on scientific proof of causality? There are times when restricting certain substances or uses—even before we have all the data or sufficient knowledge to assign a high level of significance to the results—seems like the right thing to do. Among the factors to be considered are the magnitude of the stakes; the impact of a ban or regulation; whether the product or technology in question is already in use or is expected to be introduced soon; the degree of reversibility of the outcome; whether a substitute can reasonably be found for the suspected substance or questionable technology; whether its use is voluntary or involuntary; and which parties are subject to the greatest impacts (e.g., newborns, workers, pregnant women, the elderly).

The human health implications of the environmental endocrine hypothesis are at once so profound and broadly conceived and yet so uncertain. The effects of endocrine disrupters on wildlife, although better established and in some cases demonstrated, do not provoke the same level of public concern. The high stakes and uncertainties associated with endocrine-disrupting chemicals permeating our environment are revealed in the expressions of responsibility by the scientists whose work and sentiments have shaped the public debate.

Skepticism versus the Precautionary Principle

In the case of environmental endocrine disrupters, we have a very provocative hypothesis that posits a link between a class of agricultural and industrial chemicals and health effects in animals and humans. The postulated effects in question—including infertility, cancer, sexual organ abnormalities, and behavioral and cognitive disorders in children—raise deeply emotional issues. These diseases and abnormalities play a prominent role in the worry budget of the public. Premature excitations of the public by scientists could have two undesirable outcomes. First, these messages could result in commercial losses and higher consumer costs, as well as undesirable product substitutions. Second, a high frequency of risk messages could result in a loss of public confidence with regard to claims made about potentially hazardous chemicals and products. People may become inured to the pervasiveness of risk claims since, on practical grounds, there are too many hypothetical risks to worry about.

Members of the scientific community have embraced the hypothesis of endocrine disrupters with varying degrees of personal responsibility. Their responses may be divided into four well-characterized types—perhaps even archetypes—of scientific accountability and responsibility.

Advocacy Scientists. Some scientists, who have identified themselves as supporters of the environmental endocrine hypothesis, have interpreted their personal responsibility and commitment to extend beyond their scientific work. They view their role as bifurcated between advancing the scientific knowledge base and communicating to the public, the media, and policymakers about the potential or probable risks of endocrine disrupters. These publicly visible scientists must deal with all the effects that popularization of an issue has on one's vulnerable scientific image and career. Scientists who are identified as advocates of social policy change may no longer be seen as purely disinterested in that they have made the leap between scientific evidence and what *ought* to be done. As we have seen, in his *Scientific American* review of *Our Stolen Future,* Michael Kamrin claimed that the book's advocacy undermined its scientific message because "it aims to convince readers about what ought to be rather than to explain what is or what is likely to be."[61]

The social responsibility of scientific advocates beckons them to call for controlling or prohibiting the use of some of the chemical agents in question unless their safety can be assured. For these scientists, tentative hypotheses may be good guides to social action. They believe that scientists have a duty to shape policies that reduce risk even though there are gaps in the knowledge. They are vulnerable to the charge that advancing the hypothesis is self-serving since it may mean more money for their research. Their scientific work, insofar as it advances a controversial hypothesis on the environmental etiology of disease, seems out of fashion with mainstream biomedical science, so heavily invested in genetic causation.

Theo Colborn is an interesting case in point. After a career as a pharmacist, she pursued a new professional interest in wildlife biology. She worked for a few years in the public sector as a science analyst and eventually took positions as a public interest scientist with the Conservation Foundation and the World Wildlife Fund. In these organizations, the norms of responsible science include examining the implications of knowledge for the protection of our natural resources. Scientists are on a quest to understand human causes of biological disruption and to suggest paths of amelioration. Colborn chose a nonacademic path in science and therefore was not forced to confront the classic dilemmas of scientists seeking to build a career trajectory. In those cases, the positivistic approaches of avoiding value controversies and media attention have a functional role. Colborn's strategy, on the other hand, was to build a scientific constituency for the environmental endocrine hypothesis and to use that constituency to alert policymakers and the public. Many scientists who would not themselves take on a public advocacy role were not uncomfortable providing scientific support for a colleague who had incorporated a broader concept of scientific responsibility into her approach. Their association with Colborn was based on a tacit agreement about scientific honor and integrity. She had to secure their trust by ensuring that their results and contributions to the hypothesis would not be distorted when introduced into the policy arena.

One of the techniques Colborn used to establish her credibility and to gain the trust of scientific colleagues was the development of consensus statements. After technical papers are presented at a workshop like Wingspread, the tradition followed by most organizers of scientific meetings is to publish a volume of proceedings including an introduction that summarizes the range of scientific contributions. Instead, following the global warming model, Colborn sought a consensus statement from the Wingspread participants. In pursuing this goal, she had to encourage a relationship of mutual trust among the participating scientists. To win their confidence the consensus position was designed to reflect the nuances in their interpretations of the science as well as to account for the variations in their normative beliefs. Anyone who has been part of a process through which a group of independent-minded scientists, who have assembled voluntarily, are asked to put their names to a position statement can appreciate the importance of trust and mutual respect between the facilitator and the attendees.

To win the consensus, Colborn divided the statement into several categories, covering the nature of the problem, knowledge claims, areas of uncertainty, and promising areas of research. The 1991 Wingspread consensus statement went beyond its scientific findings and recommended new testing protocols, a more comprehensive inventory of compounds in commerce, and reduced exposure to synthetic chemicals in the environment. Such regulatory recommendations are rarely outcomes of scientific meetings, unless they are mandated for that purpose by a funding agency.

A similar approach was taken at the Erice Working Session (titled "Environmental Endocrine-Disrupting Chemicals: Neural, Endocrine and Behavioral Effects") held in Sicily on November 5–10, 1995 (see Appendix B). The consensus statement framed at this working session was signed by 18 participants and included the following normative statements:

> A trivial amount of governmental resources is devoted to monitoring environmental chemicals and health effects. The public is unaware of this and believes that they are adequately protected . . . the potential risks to human health [from endocrine-disrupting chemicals] are so widespread and far-reaching that any policy based on continued ignorance of the facts would be unconscionable.
>
> Those responsible for producing man-made chemicals must assure product safety beyond a reasonable doubt. Manufacturers should be required to release the names of all chemicals used in their products with the appropriate evidence that the products pose no developmental health hazard.[62]

The consensus statement provided traditional scientists a venue to express policy judgments they believed were supported by their scientific findings. Colborn gave this group of scientists an opportunity to make trans-scientific judgments without directly exposing themselves to the media. This second group may be termed the tacit supporters.

Tacit Supporters. Some proponents of the endocrine hypothesis (i.e., those who believe it is more likely true than false) choose to limit their role to advocacy within scientific forums. They tend to shy away from the media and do not engage in writing popular essays. Without the support of these scientists, the impact of the hypothesis would have been severely limited.

Some scientists who had become converted to the idea of endocrine disrupters began rethinking their own research. A few were concerned about the professional risks of shifting their research paradigm to a hotly contested hypothesis. Citing the orthodoxy of the National Institutes of Health study sections that review proposals, one investigator commented:

> I was funded for 13 years at NSF [National Science Foundation], then USDA [U.S. Department of Agriculture] and NIH [National Institutes of Health]. I was studying reproductive success and organ development as it impacts natural populations. In 1990, I submitted a grant [proposal] to NIH on endocrine disrupters that went to the toxicology study section. . . . The study section replied to my proposal that they knew of no evidence that hormones influence development and why would I do these silly experiments.
>
> We have nowhere to go with our proposals. If we mention toxicology to a basic science panel, they are not interested. If your proposal goes to a toxicology panel, they don't know enough to read it. So, no funding. We wrote grant proposals that had no place to go.[63]

But not everyone in this category of scientific supporters believed that the preliminary data required public action. Niels Skakkebaek is known as a strong proponent for the hypothesis that chemical exposures are causing a decline in the quality and density of human sperm. It was because of his work that the Danish government supported an environmental study of male reproductive issues. Yet when asked by the environmental ministry of Denmark whether there was enough evidence to pass legislation regulating chemical endocrine disrupters, Skakkebaek was not prepared to suggest legislative reforms. We must be cautious in our policy proposals, he said. "I am not educated enough in biochemistry to know whether it would be the right thing to do. People are so greedy that they will probably not accept a decline in living standards very much. If you prohibit something, something else might come up and who could say the replacement would be better?" Skakkebaek thought it would be wrong for him to be an advocate for policy. "I could [direct] my energy much better by telling people there is a problem and trying to provide more information."[64] Scientists must be willing to accept the fact that their deepest scientific convictions might be false. For the group I call tacit supporters, the vulnerability of their belief reinforces their resolve to push on with their research yet casts a conservative cloak over their policy outlook.

Critical Skeptics. A third category of scientists are the critics of the hypothesis, who may express their skepticism in two ways. First, as scientists, they serve as the critical counterpoint to proponents of the endocrine hypothesis. Their critiques eventually lead to the strengthening, weakening, or falsification of the postulated effects. Second, some skeptics believe they have a public responsibility beyond their commitment to science to prevent the "Chicken Little" syndrome of acting prematurely and raising many false alarms. They support a value system that implores us to act when and only when there is sufficient evidence and not on the basis of tentative or speculative hypotheses. They reject the precautionary principle and would rather see policy built on "sound science," science that proves a cause-effect connection. Typically, their values and behavior raise questions among their detractors about any associations they might have with those entities within society that would be financially harmed by taking precautionary actions against chemicals.

One of the most publicly visible scientific skeptics of the endocrine hypothesis, Stephen Safe, published his first major critique in the journal *Environmental Health Perspectives* just a month before *Our Stolen Future* was released. The article was titled "Environmental and Dietary Estrogens and Human Health: Is There a Problem?"[65] He has also written a critique of the putative association between xenoestrogens and breast cancer, one of the subhypotheses under the theory of endocrine disrupters.[66] Safe disclosed that financial support for his work on antiestrogens came from the NIEHS, that his research on breast cancer and xenoestrogens was supported by the National Institutes of Health, and that his research on dietary and environmental estrogens was funded in part by the Chemical Manufacturers Association, an industry organization that has lobbied against chemical regulations.[67] Because of his support from the association (amounting to about $150,000 per year over three years for the study of weak estrogens), Safe's objectivity toward the hypothesis has been questioned. However, this factor did not prevent the National Academy of Sciences from appointing Safe to its study panel, the Committee on Hormone-Related Toxicants in the Environment. The academy also appointed to the study panel scientists who expressed strong support for the hypothesis.

At an earlier stage in his career, Safe's scientific research had not always been viewed as friendly to industry. In 1969–70, while he was a research officer at the National Research Council of Canada and based at the University of Guelph, Safe was investigating the chemistry and biochemistry of organochlorines. He published papers on the metabolism and photodegradation of PCBs and coauthored one book on mass spectrometry of pesticides and pollutants and another on the chemistry of PCBs. During the 1970s and 1980s, Safe became involved in controversies surrounding organohalogen contamination in the Great Lakes. During occasional interviews with the media he recommended that women not breastfeed their infants because of the levels of organohalogens in their breast milk. He also testified on behalf of farmers in Michigan whose cows were exposed to low levels of polybrominated biphenyls (PBBs) during the state's massive PBB contamination episode, when the industrial chemical appeared in animal feed.[68]

The issue of credibility and funding sources was also raised by the critics of the hypothesis. A background paper on *Our Stolen Future* released by the American Council on Science and Health cited support for the book from six foundations (W. Alton Jones, Joyce, C. S. Mott, Pew, Winslow, and Johnson) and then stated that "these foundations have records of funding organizations and individuals who take radical environmentalist positions inconsistent with mainstream scientific perspectives on either risk assessment or the public health benefits associated with technology."[69] Although these foundations did support Colborn's work over the years, there was no direct support from these sources for the writing of the book itself. That came from the publisher's advance.

Safe's critique of the hypothesis hammered at the inconclusiveness of the data used to support certain claims, particularly with respect to the increased incidence of breast cancer and decreased sperm counts. He indicated that no one has yet disentangled the role of dietary endocrine modulators (bioflavonoids) from that of environmental endocrine disrupters (organochlorine pesticides).[70] His critique of the human health component of the hypothesis is based on his view that its proponents have failed to (1) present definitive human health data, (2) address the relationship between industrial compounds and naturally occurring estrogenic compounds, or (3) establish the associations or health trends claimed, such as decreased sperm count or the link between breast cancer and high serum levels of organochlorine pollutants.

In a letter submitted to the journal *Environmental Health Perspectives,* Safe called upon scientists to be more skeptical about the effects attributed to synthetic chemicals. His words resonated with the philosophy of Karl Popper, who argued that only those hypotheses that can stand up to the challenge of falsification are worthy of our confidence: "So long as a theory withstands detailed and severe tests and is not superseded by another theory in the course of scientific progress, we may say that it has 'proved its mettle' or that it is 'corroborated.'"[71] Safe wrote:

> Environmental studies show a possible linkage between exposure to hormone mimics and possible adverse effects in fish and wildlife. It has been hypothesized that these hormone mimics may be causing adverse effects in humans; unfortunately, scientific hypotheses are often treated in the press as scientific facts. In contrast, as scientists, we tend to be questioning and skeptical of hypotheses in the absence of data. . . .
>
> The normal human diet contains diverse endocrine disrupter and hormone mimics which in themselves may cause adverse effects. Therefore, consideration of environmental hormone mimics must take into account other background exposures to these same types of compounds in the human diet. . . . My review article concluded that based on current data, adverse human impacts of industrial estrogens were unlikely.[72]

Safe's critique raises the questions: Is it responsible to advocate the regulation or prohibition of chemicals without definitive proof of causal action? Should we not consider alternative explanations before we rush to judgment? Does one behave responsibly in the policy arena by waiting for definitive evidence or by relying on circumstantial evidence? The authors of *Our Stolen Future* speak of the "preponderance of evidence" as a basis for responsible action. This dilemma is complicated by another factor, cited by Hertsgaard in his essay in the *New York Times Book Review:* "The authors of *Our Stolen Future* acknowledge they do not have airtight proof that hormone-disrupting chemicals are to blame for reproductive problems. Such proof, they point out, is beyond anyone's power to demonstrate in a world awash in man-made chemicals. Humans are now inescapably exposed to so many chemicals that it is effectively impossible to prove a cause-and-effect connection between a given compound and a corresponding malady; individual substances cannot be sorted out from the overall mix."[73]

If this conclusion is correct and the causality of chemical effects on humans is unknowable by the conditions of exposure, then some alternative norms of responsible social behavior must emerge to replace the traditional scientific model of hypothesis validation. When public health is at stake, the authors of *Our Stolen Future* have proposed a methodology for responsible action: "One assesses the totality of the information in the light of epidemiological criteria for causality, such as whether the exposure precedes the effect, whether there is consistent association between a contaminant and damage, and whether the association is possible in light of the current understanding of biological mechanisms. But this real-world environmental detective work comes to judgment based on 'the weight of evidence' rather than on scientific ideals of proof that are more appropriate to controlled laboratory experiments and the practice of science than to problem solving and protecting public health in the real world."[74]

What standards and norms underlie the use of "weight of evidence" criteria for making policy judgments? Many public health decisions—including those in response to urea formaldehyde, radon, lead, and asbestos—were made prior to definitive proof of dose-response effects and exposure assessments for humans. Responsible public actions could build on such cases, in which prudent avoidance, the precautionary principle, and "weight of evidence" are incorporated into social decisions.

Silent Skeptics. Thomas Kuhn argued that scientists are not likely to switch paradigms by the force of evidence. He described the reception to new ideas as more like a conversion process than a logical or rational progression of thought. Skepticism is built into the scientific agenda. Some would say it is the raison d'être of science. As such it is not surprising that many scientists will view the endocrine hypothesis as a speculative and bold theory that fails to make the causal case for human effects. There are several factors about the science that nurture this skepticism.

First, the endocrine hypothesis covers a wide range of effects. Its evidence comes from the methods and results of many disciplines. Scientists generally shy away from constructing grand-scale hypotheses that are transdisciplinary, and most scholarly journals do not publish grand, theory-building articles. The antipathy toward overarching theories is built into the very culture of science, which weaves its tapestry of knowledge into a patchwork quilt of highly differentiated subdisciplines.

Second, there is no body of definitive data linking ambient exposure levels of environmental endocrine disrupters to a human illness or reproductive abnormality. Scientists who are dubious about extrapolating the results of animal studies to humans, of in vitro effects to in vivo effects, or of high-dose exposures to low-dose exposures will bring that skepticism to bear on any hypothesis for which the evidentiary support comes from one or more of these indirect sources.

Whether one holds to a Kuhnian interpretation of scientific change or a rational model, new hypotheses in science do not have an easy time of it. Scientists who have not expressed support for the endocrine hypothesis fall into two categories: those who feel it is ill conceived and does not merit the attention of the scientific world, and those who feel the case has been made for plausibility and that the hypothesis is indeed worthy of time and resources from the national research agenda. The silent skeptics will usually play a role in the background as reviewers of papers and grant proposals—as gatekeepers for journal editors. Their role in the landscape of science is to ensure that any specific finding is not hastily and uncritically converted into support for the central tenet without consideration of other interpretations. Sometimes tacit skepticism has other roots. Steingraber, commenting on Rachel Carson's observations about silence among scientists who knew better, noted: "The third kind of silence that fascinated Carson was the hushed complicity of many individual scientists who were aware of—if not directly involved in documenting—the hazards created by chemical assaults on the natural world. While dutifully publishing their research, most were reluctant about speaking out publicly, and some refused Carson's requests for more information. Writing in *Silent Spring,* Carson acknowledged the constant threat of defunding that hushed many government scientists."[75]

The release of *Our Stolen Future* has forced many scientists to take notice of the central thesis of the environmental endocrine hypothesis. Their expressions of skepticism in reviews of the book were a sign that the hypothesis had emerged from obscurity to high visibility. Within two years the reception of the hypothesis among scientists had changed significantly. By 1996 scientists from academia, government, and industry were being invited to scientific meetings where environmental endocrine disrupters were an accepted part of the discussion. A new scientific agenda had emerged that included studies of the effects of endocrine disrupters on humans and wildlife, methods of measuring exposure, and assays for screening and testing chemicals for endocrine-disrupting effects. But as we shall see, one of the first major efforts to lay a foundation for risk assessment turned into a political minefield.

The Synergism Backlash

It is generally understood from studies of human serum and tissue samples that our bodies are host to a cocktail of trace amounts of industrial chemicals that did not exist prior to the twentieth century. Many of these chemicals can persist for years in body fat and blood serum. On occasion, they may be transported from the body of a pregnant woman through her placenta to the developing fetus in sufficient concentration to damage the offspring. Or chemicals metabolized in a woman's body may be mobilized during lactation and delivered via breast milk to her infant child.

Among the most challenging questions in human toxicology are: What is the combined effect of these postindustrial chemical cocktails on human biology? Can the interaction or the accumulated amounts of subtoxic quantities of individual chemicals result in toxic effects? The environmental endocrine hypothesis provided a new way of framing this question: Can endocrine-disrupting chemicals be acting synergistically? That is, could a mixture of two or more estrogenic chemicals produce an effect greater than the simple additive contributions of its components? *Synergism* is the short form for the expression "the whole is greater than the sum of its parts." But one scientist who made an effort to understand whether synergism plays a role in the actions of endocrine-disrupting compounds stepped on a scientific land mine.

John McLachlan left the NIEHS in 1995 to take a position at Tulane University. Soon thereafter, he led a team of researchers in an investigation of whether the pesticides endosulfan, dieldrin, toxaphene, and chlordane were synergistic with respect to the inducement of estrogenic activity. Each of these pesticides is capable of binding with the human estrogen receptor and producing estrogen-associated responses, albeit at low concentrations. The research team developed a cell culture model to test the effects of combinations of these pesticides on certain estrogen responses.

The Tulane investigators used a yeast cell containing a human estrogen receptor. Under the proper conditions, the yeast cell's genetic system can be made to express (carry out the synthesis of) foreign genes. Chemicals that bind to the estrogen receptor can induce transcriptional activity (activate protein synthesis) within the yeast's genome. The researchers measured the synthesis of the protein β-galactosidase (β-Gal)—an estrogen-active product—based on chemical exposures to the four pesticides, alone and then in pairs. They reported that "a combination of any two of those chemicals produced a synergistic increase in β-Gal activity as compared with similar quantities of the individual compounds."[76] Had the synergism they reported been slight or even modest, it might not have created the impact it had. But the Tulane scientists reported estrogen potency in mixtures up to 1,600 times what they measured for the same quantities of individual agents. Although others had previously reported synergistic effects of environmental chemicals, including pesticides, these new results suggested that the synergistic effects of estrogenic chemicals had been significantly underestimated. They called into question the reliability of all health standards based on low-exposure assessment of pesticide residues.

A paper reporting the results of the Tulane study was submitted to the journal *Science* in February 1996, accepted for publication in May, and published in the June 7, 1996, issue under the title "Synergistic Activation of Estrogen Receptor with Combinations of Environmental Chemicals."[77]

The interest of chemical producers in these results was immediate because they presented the threat of more aggressive regulation. Congress had just revamped the laws on pesticide residues in food to end zero tolerance for carcinogens but raise the standards for acceptable risks—a change widely sought by the chemical industry. Trade groups promptly dedicated funds to verify the synergism results of the Tulane scientists. Yet two sets of experiments that sought to replicate the Tulane experiment failed to demonstrate synergism. The results were reported in *Science* and the journal *Endocrinology* between January and April 1997. A news release from the Chemical Industry Institute of Toxicology on May 30 reported the failure to replicate the Tulane results: "Scientists from four major research organizations, using 10 different test systems, have demonstrated that low concentrations of mixtures of weakly estrogenic chemicals do not produce greater effects than predicted from the study of individual chemicals. The new findings allay concern over low-level exposures to either synthetic or naturally-occurring chemicals that have weak estrogenic activity, acting like the major female sex hormone, estrogen."[78]

With both U.S. and European teams of scientists reporting that they could not replicate the Tulane synergism results,[79] McLachlan's group was in the uneasy position of having either to replicate the study themselves in a demonstration to their colleagues or, on failing to do so, to comply with established scientific tradition and retract their results. The gossip around the industrial community and among the skeptics of the hypothesis was that McLachlan was mistaken about synergism, and they were poised to rub the noses of those who supported the general hypothesis in this mistake. In the July 25, 1997, issue of *Science,* McLachlan, writing as corresponding author of the paper, published a retraction letter citing his team's inability to replicate the paper's principal results and concluding that the design of the original experiment had been flawed: "Whatever merit this publication [the June 7, 1996, *Science* paper] contained, and despite the enthusiasm it generated, it is clear that any conclusions drawn from the paper must be suspended until such time, if ever, [that] the data can be substantiated. . . . It seems evident there must have been a fundamental flaw in the design of our original experiment."[80]

Retracted studies, although not commonplace, are also not that rare in scientific publications. In fact, just several weeks before the McLachlan retraction *Science* had published another letter retracting the results of a biological experiment published in April of the same year because the results were not consistent with a replicated experiment.[81]

But few retractions receive the notoriety or produce the public response that resulted from that of the synergism paper. The finding of pesticide synergism had been publicized widely through environmental networks on the Internet and in popular magazines. Within months the results had been cited in the popular book *Toxic Deception.*[82] The published retraction was welcomed by the chemical industry and used opportunistically by opponents and skeptics of the endocrine hypothesis.

None of the major national newspapers carried the story of the retraction when it was first published in *Science.* Nearly a month later the *Washington Post* ran a story questioning the short time that had elapsed between publication and retraction and reporting that Tulane University was conducting an internal inquiry to investigate whether proper laboratory practices had been followed—a euphemism for a scientific misconduct investigation, which was subsequently not corroborated by officials at the university.[83] On the heels of the *Washington Post* report, on August 20 the *Wall Street Journal* ran an editorial by Stephen Safe, the Texas A&M University toxicologist and critic of the hypothesis, under the lead "Another Enviro-scare Debunked." Safe was also a member of the National Academy of Sciences panel on endocrine disrupters. Ironically, the academy requests that panel members avoid political and policy statements on the subject of their panel study during the tenure of their service. Safe's editorial stated that "the best science now points to the conclusion that xenoestrogens and related compounds are less harmful than had been suggested." Safe argued in his commentary that xenoestrogens are not a cause of breast cancer or sperm decline as some would have us believe. He wrote that estrogenic chemicals do not behave synergistically and that claims about their effects on neurodevelopment in children have not been proven. He asked rhetorically whether Congress had acted precipitately in enacting legislation requiring the screening of endocrine-disrupting chemicals.[84]

The chemical industry exploited the synergism retraction to promote its opposition to the general theory linking endocrine-disrupting chemicals to human abnormalities and health effects. An editorial writer for the

Detroit News echoed the industrial reaction when she wrote in the *Wall Street Journal* that "the endocrine apocalypse has been canceled," comparing it to other alleged false alarms like those prompted by Alar, asbestos in schools, cyclamates, dioxin, and electromagnetic fields.[85]

Another, more sobering, viewpoint on the retraction was offered by the editors of *Environmental Health Perspectives,* the journal that has devoted more pages than any other to endocrine disrupters. In the August 1997 issue, the editors noted that the speedy retraction of the synergism results meant that they would not serve as a building block for other scientific studies published in that journal. The editors sought to allay the concerns of readers that there would be a domino effect from the retraction, in which other articles published in *Environmental Health Perspectives* reporting results based on the synergism data would have to be retracted. They also pointed out that, in studies of PCBs and other pesticides, synergy and antagonism of chemical mixtures had been demonstrated at certain levels. The editors praised McLachlan and his colleagues for their scientific integrity in responding so quickly to the skepticism of the scientific community and to the anomalies that had been pointed out in their results:

> The history of science included investigators who have clung to untenable positions maintaining their delusions to the bitter end, and most withdrawals were forced only after evidence had been offered suggesting some form of misconduct. The voluntary withdrawal of a scientific paper because the data cannot be substantiated is a rare event. While these actions have been very painful for all involved, they are an essential part of the process that makes science a unique human enterprise. In science, the fallibility of human involvement is minimized over time by observing and re-observing, testing and retesting. Data that do not support the consensual reality of science are replaced or quickly forgotten. In facilitating this process by withdrawing their paper, McLachlan and his colleagues have served science appropriately and well.[86]

The retraction of the synergism results did not weaken the resolve of EPA scientists to pursue the cross-reactive and additive effects of endocrine disrupters. Two months after the published retraction, James McKinney of the EPA, in an editorial in *Environmental Health Perspectives,* argued for the need to expand our thinking about the possible human health risks from exposure to chemical mixtures: "Although many/most of these chemicals may function as imperfect hormones with relatively low potencies, we have not begun to understand what the potential adverse effects are of being exposed continuously to complex mixtures of chemicals with varying abilities to affect multiple signaling pathways both singly and interactively."[87] The synergism retraction had no appreciable effect on the EPA's legislatively mandated process for developing a screening and testing program for endocrine disrupters.

Industry Reactions and Counteractions

When an industrial sector as large and powerful as the chemical industry is confronted with scientific hypotheses that threaten its profits through new regulations or that could foster consumer or occupational liability suits, one can expect strong anticipatory self-protective responses. Many of the standard industrial defenses were put into place within a few years after the first Wingspread Work Session on endocrine disrupters in 1991. Industrial trade organizations such as the Chemical Manufacturers Association and the European Chemical Industrial Association set aside funds to investigate scientific claims linking their products to human health effects and to prepare for a counteroffensive against claims they believed were weak or vulnerable. Academic scientists who expressed skepticism toward the general hypothesis were sought out and funded to engage in studies that the chemical companies believed would exonerate their products—or at least weaken the grounds for implicating their chemicals as endocrine disrupters.

Poised for a new battle in the long war against synthetic organic chemicals, the chemical industry mobilized quickly to counteract the publicity accorded *Our Stolen Future* by soliciting critical reviews. And through its vast public relations network, the industry exploited any results that were not consistent with the idea that low doses of synthetic compounds could adversely affect wildlife or humans. Industry-funded organizations such as the American Council on Science and Health issued reports and news releases arguing there was no credible scientific evidence to support the claims of the environmental endocrine hypothesis and that the arguments underlying it were ideological. Internet sites supported by chemical companies sought to neutralize the public concerns about endocrine disrupters that were rapidly being incorporated into environmental activist agendas.

Starting around 1996, the Chemical Industry Institute of Toxicology (a respected research institute that is heavily supported by major chemical companies) was spending about $1.5 million annually investigating claims of proponents of the environmental endocrine hypothesis. Some of the institute's work is published in the open literature and subject to peer review. Findings that implicate products may not be published as quickly as findings that exonerate them, and the institute's threshold of skepticism for accepting chemical risks may be at the high end. But when the evidence is unambiguous and incontrovertible, the institute has a reputation for reporting the findings. For example, Paul Foster, manager of the institute's Endocrine, Reproductive and Developmental Toxicology Program, wrote an article in the organization's journal that reviewed toxicological studies of one of the chemical products in the family of phthalates: dibutylphthalate (DBP), a chemical used extensively in plastics and solvents. In a rhetorical style that bristled with understatement and cautions, Foster asserted: "The results of this study provide some support for the hypothesis, first proposed by R. M. Sharpe and N. E. Skakkebaek, that exposure to endocrine-active chemicals in utero during critical periods can result in a number of male reproductive deficits in adulthood." Moreover, he noted that the lowest observed adverse effect level of 65 milligrams per kilogram per day based on rat studies might be exceeded fourfold in infant exposures for worst-case scenarios.[88] This was not a result the chemical industry was waiting to hear.

The strategy of the chemical industry has been to fund its own research while investing in public relations that cast a positive light on the industry. Nevertheless, industry reaction to the periodic news reports on the chemical risks of endocrine disrupters has been measured. There have been no strident counterattacks, only positive advertisements during prime-time broadcasts, through 1997 and continuing into 1999, extolling the benefits of plastics. The advertisements feature images of small children in bicycle helmets, adults with biomedical life-saving devices, and newborns in plastic incubators, all accompanied by a comforting voice assuring viewers that "plastics make it possible." The campaign is reminiscent of the 1950s theme "Better Living Through Chemistry." These promotional ads are designed to create a positive public image of plastics—an image that was threatened by disclosures that certain compounds used in the manufacture of one category of plastics, polycarbonates, exhibited estrogenic effects in animal studies. Similar broadcasts that show toddlers romping across a newly herbicided lawn are intended to remake the public's image of agrichemicals.

The scientific work largely responsible for the industry discomfort was carried out by developmental biologist Frederick vom Saal of the University of Missouri. As we have seen, until vom Saal met Theo Colborn he had not considered that his research on intrauterine positioning and hormonal effects on mice might have environmental significance. Subsequent to Wingspread I, vom Saal adapted his mouse model for studying the effects on development of endocrine disrupters.

To investigate the effects of low doses of endocrine disrupters on the offspring of pregnant mice, he exposed the pregnant mice to bisphenol-A, the industrial name of a compound (with at least 18 synonyms in the chemical nomenclature) widely used in the manufacture of resins and polycarbonate flasks; the chemical's estrogenicity had been identified in the 1930s.[89] From his own experiments, and other published studies both in vitro (in cell cultures) and in vivo (in rats and mice), vom Saal concluded that bisphenol-A is bioactive within the range of human exposure: "Taken together, those findings suggest that an alteration in the course of development of reproductive organs could also occur in human fetuses carried by pregnant women who consume amounts of bisphenol-A found in canned products or foods heated in polycarbonate containers."[90]

Chemical companies considered such animal studies implicating chemicals at very low doses as highly destabilizing to the industry. Vom Saal began to receive inquiries about his work. At first they seemed innocent enough. There were requests for reprints and detailed information about his procedures. He became a little suspicious when one chemical company sent a scientist to his laboratory to learn how to work with his animal system. His suspicions were soon borne out, as he discovered there was another agenda: "Dow Chemical sent a guy down here and he said can we arrive at a mutually beneficial outcome, where you don't publish this work on bisphenol-A until the chemical industry has replicated your study, and approval for publication was received by all the plastic manufacturers. I was stunned." When the company representative failed to bring vom Saal on board with the "carrot," he resorted to challenging the scientist or attempting to shake his confidence in the results, using such expressions as "we are very disturbed by your work" and "how can you publish something when you don't know all the mechanisms underlying what you are doing?"[91]

Personally offended by the solicitations of this industry messenger, vom Saal and his collaborator Wade Welshons fired off an irate letter to the Society of the Plastics Industry expressing their belief that this behavior on the part of such representatives was totally inappropriate:

> We were surprised that Dr. ——— came here with the task of asking us to withhold publication of a paper on bisphenol-A which is in press in the *Journal of Toxicology and Industrial Health.* Dr. vom Saal had made

> available an initial draft of this paper to everyone from the Plastics Industry who attended a meeting at MPI Research during his visit there in February 1997. In this paper we report effects of fetal exposure to bisphenol-A on sperm production, and both epididymidal and preputial gland weights. Dr. —— began by stating that it was the hope of Dow Chemical that there should be "some mutually beneficial outcome" as a result of withdrawing the paper and postponing publication until a replicate study of effects of bisphenol-A in mice to be conducted by MPI Research has been completed and *approved for publication.* We stated that we would not withdraw the paper unless Dr. ——— could provide a scientific basis as to why it should not be published.

They reminded the society of "recent scandals in the pharmaceutical industry where pharmaceutical companies have attempted to intimidate university scientists who have found results that raised questions about a product. This type of response," they indicated, "can seriously undermine the credibility of the companies that manufacture a product that becomes the subject of controversy."[92]

The Society of the Plastics Industry continued to support research to refute vom Saal. In the fall of 1998, the society held a press conference in Washington, D.C., releasing an unpublished study allegedly refuting vom Saal's finding that bisphenol-A caused abnormal prostate growth and low sperm counts in mice.[93] Vom Saal had not received any financial support from the chemical industry for his work on bisphenol-A—in contrast to other notable cases of industrial influence on research, in which companies threatened the most by the results of the studies had exerted financial or legal leverage over the researcher.[94]

In the post-Alar, post-Bhopal period, some corporations have learned how to react to public criticism, how to project a "green" image, and how to avoid personal attacks on their critics that could backfire. Some industry-oriented magazines have advised companies to take positive steps to reduce production of endocrine disrupters, even if the environmental endocrine hypothesis proves to be nothing more than a hypothesis.

The May 6, 1996, issue of *Chemistry & Industry* carried the following advice for manufacturers:

> If the industry believes its own tired rhetoric on sustainable development, self-regulation and better public relations, it will have to make more effort than it did with CFCs to respond in a credible fashion. For one thing, it is crucial to avoid even inadvertently suggesting

> that the "no evidence of harm" somehow equals "evidence of no harm." Equating the two, if only by implication, is just as misleading as the environmentalists' exaggerations.
>
> There is a better way for industry: get ahead of the issue, not behind it. First, start looking now for the most cost-effective first steps to eliminate or reduce exposure to some putative hormone disrupters, and take these as soon as they are identified. . . . Second, find new people to run the industry's public relations in Europe. Response to this issue is too little, too late, and too unsophisticated. . . . Oestrogen mimics have been bubbling inexorably to the top of the public agenda for at least four years now, plenty of time to cobble together more than some excuses about the difficulty of coordinating an industry view. . . . Third, put substantial funding and a sense of urgency into answering all those questions that stretch across the gaps in data.[95]

The value issues associated with the endocrine disruption hypothesis are emblematic of many cases in which science informs public decisions. When should concerns over environmental hazards be transformed into policy? How should scientists respond to the media's bottom-line approach to environmental reporting? What are the tradeoffs for scientists who become advocates for a cause?

No generalizable patterns of ethical responsibility emerge from this episode, only scientific roles, contextual factors, and idiosyncratic situations. Some scientists are prepared to make a personal decision on the basis of preliminary circumstantial evidence. A speculative hypothesis may be worth acting on because if you are wrong little is lost, and if you are correct a great deal is gained. A stronger position is found in the precautionary principle, whereby one should minimize regret even when the evidence is weak but suggestive and even when the cost of social restraint and industrial compliance is high. One scientist confided that she is not sure that the environmental endocrine hypothesis is correct, but just in case it is, she does not microwave her food in plastic containers. However, not all conjectures lend themselves to such a risk-benefit calculation. If we were to act on the conjecture that mercury amalgams are dangerous, then we might be inclined to have all our mercury fillings removed, which could be expensive and potentially more dangerous than leaving them in place. Or consider the hypothesis that exposure to high-intensity electromagnetic fields may cause cancer. For household electromagnetic fields, there is not much evidence to go on. However, consumers have been warned against sleeping with electric blankets. Perhaps that is not a significant cost for the benefit of preventing illness if the postulated connection between electromagnetic fields and health effects eventually proves correct.

Within the domain of second-order science, namely science applied to public problems, we have a frontier ethic. Almost anything goes. Scientists position themselves either as protectors of the canons of their discipline from challenge by renegades or as critics of the orthodoxy who offer a new perspective on environmental illness. One principle to be considered might be: We do not accept a conjecture on environmental risk as true until the evidence is in. That is the scientific way. The scientist's behavior toward a conjecture is guided by the idea that truth must be demonstrated. However, in some instances the political agenda calls for arousing public sentiment about a *potential* risk, for that may be the only way to mobilize funds for research that may reduce the margins of uncertainty around the hypothesis.

The following story illustrates some of these points. I had the opportunity to review a draft of an article on xenoestrogens released from the lacquer coating in food cans. The lead scientist was a biologist at the Laboratory of Medical Research and Tumor Biology at the University of Granada. I communicated to an intermediary scientist I knew at the Tufts University School of Medicine that I thought the paper could be sent to a top-tier international journal of general scope, not because of any pathbreaking science per se but because the issue of xenoestrogens has become quite important and the paper's results would be of interest to a broad sector of the science and policy communities. I suggested *Nature* for several reasons, including the fact that it had been relatively open-minded on issues of toxic chemicals and the environment. The Washington, D.C., editor of *Nature* was prepared to pouch the paper to London with a forwarding note introducing it to the home office editor.

The paper reported on the testing of canned foods for the release of bisphenol-A, a material that is used in the lacquer coating of the cans. The authors stated: "The data reported in this paper strongly suggest that some foods preserved in lacquer-coated cans acquire estrogenic activity." The editor of *Nature* sent the paper to two reviewers, a cancer epidemiologist and a reproductive biologist specializing in estrogen effects. They did not submit written reviews to the *Nature* editor. In communicating with the author, the editor paraphrased the reviewers' oral comments. His rejection letter stated: "Because there is as yet no definitive evidence that environmental oestrogens have an adverse effect on health . . . your results at this stage are more appropriate for the specialist literature than for

Nature." The paper was published in *Environmental Health Perspectives* several months later under the title "Xenoestrogens Released from Lacquer Coatings in Food Cans."[96]

As expected, its publication was barely noticed by the media. The reviewers for *Nature* had shielded the journal from publishing results that related to a widely acknowledged hypothesis, applying criteria that could easily have prevented the journal from publishing early work on ozone depletion, global warming, or the toxicity of many substances for which there is also no *definitive* evidence of human disease. In this case the gatekeepers ruled that the endocrine hypothesis was not ready for prime time, notwithstanding the fact that for several decades scientists have been accumulating evidence that compounds introduced into the environment are capable of disrupting the endocrine system of animals and humans. No credible scientific spokesperson has thus far claimed there is definitive evidence that xenoestrogens cause human health problems. But many voices from many continents and many disciplines continue to assert the likelihood that what has been observed in wildlife and laboratory animals is probably also occurring in humans.

The next chapter looks at some fundamental questions pertaining to public policy and scientific uncertainty and traces the beginnings of a regulatory response to endocrine disrupters.

CHAPTER

4

The Policy Conundrum

Science and policy are two symbiotic cultures that would prefer to live independently but are resigned to an uneasy, if reciprocal, coexistence. Scientists are forever asking for additional studies and the public funds to carry them out. Policymakers are always seeking to meet the public's demands for increased safety with limited knowledge, short timetables, and declining budgets. Science proceeds slowly and incrementally. Policy may change erratically and apparently illogically. And when it comes to environmental causes of disease, science can rarely provide a crucial experiment, unambiguous data, or convergence from all the experimental domains that are relevant to the policy issue. On the one hand, scientists want to keep politicians out of their work. On the other hand, politics is often the impetus behind the funding of science and the crafting of policy.

This chapter explores the intersection of science and policy on the question, What regulatory considerations should be given to endocrine-disrupting chemicals? As this book went to press, the question had been only partially resolved within the American and international policy sectors. However, after 10 years, the environmental endocrine hypothesis has been responsible for new research initiatives and a legislative commitment to discover which chemicals in current use have the potential to cause endocrine disruption in wildlife and humans. This is a critical step on the path toward the management of risks associated with exposure to the chemicals of modern life.

Hypothesis Formation and Public Policy

Scientists do not frequently organize themselves to advance a public health agenda or an environmental risk hypothesis. There have, however, been some notable exceptions. In the 1950s, physicists, biologists, and chemists (led by Nobelist Linus Pauling) alerted the public that atmospheric testing of atomic weapons posed significant health risks to populations hundreds, even thousands, of miles from the test site. Pauling organized a petition signed by over 9,000 scientists and submitted to the secretary general of the United Nations that requested a ban on atmospheric testing of nuclear weapons.[1] Scientists sought to alert citizens to the health risks of such radioactive fallout products as carbon-14 and strontium-90; the latter was found in the milk of lactating mothers, as well as the bones and teeth of young children.[2]

A second example was the Asilomar meeting of 1975, an international conference convened by biologists to discuss the potential hazards of recombinant DNA research.[3] The meeting resulted in guidelines for genetic engineering research issued by the National Institutes of Health and the establishment of a governmental oversight body.

In another case, a group of atmospheric and environmental scientists, having first posited a relationship between chlorofluorocarbons and the depletion of atmospheric ozone, organized themselves as a force for change in public policy.[4] One of the outcomes of this largely science-directed initiative was the signing of the Montreal Protocol in 1987, which included agreements on tighter quotas on the amount of chlorofluorocarbons the signatory nations would be allowed to produce in any given year as well as heavy taxes on the continued use of the chemicals.[5]

In each of these cases, the scientific findings were only one of several factors that carried the hypothesis from scientific circles to the public policy arena, even while there was uncertainty over the risks—and in some cases no concrete evidence that there was a risk. The scientists who engaged in a public dialogue played multiple roles as generators, interpreters, purveyors of knowledge, and, for some, advocates for policy change.

There are many other instances in which scientists at America's premier institutions trailed rather than led the advance of a risk hypothesis to a national policy forum. A recent example is the Alar episode. The chemical Alar (the trade name of daminozide) is a growth regulator once commonly used in preharvest fruits to establish uniformity in ripening and color. Alar was removed from use by farmers when the public stopped buying fruit treated with the chemical after a vigorous media exposé of its potential cancer-causing properties.[6] The issue was brought to the attention of the media and eventually the public through the publication of a report by the Natural Resources Defense Council on the cancer risks Alar and other pesticides posed to children (see Chapter 3).[7] Years after the consumer boycott, scientists and policy analysts continue to debate whether Alar should have been removed from the market, suggesting that there are significantly different action thresholds within science and public policy for reaching closure on the potential risks of chemical exposures.

In the course of scientific inquiry it is not unusual for the validity of a hypothesis to remain in question for many years before it is either rejected or incorporated into the canons of established knowledge. After 50 years of skepticism, scientists eventually adopted the hypothesis that plants can obtain nitrogen from the air.[8] Today, a discussion of nitrogen fixation by bacteria that reside in the root nodules of plants appears routinely in basic biology texts, ending years of skepticism that plants could only obtain nitrogen from soils. It took the scientific community some 16 years to reach a consensus that chlorofluorocarbons were responsible for the breakdown of the protective ozone shield, after a detectable "hole" was identified over Antarctica.[9]

Typically, one can chart a gestation period for the adoption or rejection of unorthodox hypotheses. There is a pre-establishment phase, in which a hypothesis may be advanced in the so-called gray literature (unpublished reports and non-peer-reviewed journals) or in popular science publications without having been published in mainstream science journals. Once a hypothesis reaches the established science literature, a small group of its supporters makes the case by advancing the thesis in symposia, by entering into debates in the journals, and by seeking additional forms of evidentiary support. During this transitional period, the antagonists and protagonists create the tension and self-criticism that provide quality control in science. The emergence of adversarial camps is central to ensuring that a hypothesis undergoes careful scrutiny before its final disposition. Criticism is indispensable to the growth and certification of knowledge.

The final phase in the life cycle of a hypothesis can take several forms. The status of the hypothesis may emerge clearly and definitively to the vast majority of the scientific community. Usually this occurs after new evidence appears in the form of a crucial experiment or critical discovery that plays a deciding role in the acceptance or rejection of the hypothesis. At other times, the fate of a hypothesis is determined slowly and incrementally. Proponents grow in numbers until a functional consensus emerges among leaders in the field. As in the Kuhnian paradigm shift, certain hypotheses finally are accepted into the canons of science when older skeptics die, retire, or are outnumbered. Alternatively, an unpopular hypothesis may age along with its few dedicated proponents and slowly disappear from the scientific literature.

There is an important distinction between hypotheses of pure science, such as the nitrogen fixation hypothesis, which is now an established fact, and hypotheses that have important social or environmental implications, such as the ozone depletion hypothesis. For most of its gestation period, the nitrogen fixation hypothesis was of interest exclusively to scientists. In contrast, from the outset, the ozone depletion hypothesis had important public health and environmental policy implications. A reduction in atmospheric ozone could result in global radiation imbalances that might endanger the survival of species or at the very least cause an increased incidence of skin cancer. When a scientific hypothesis has important implications for public policy, the time needed to acquire conclusive evidence of its validity takes on a special significance. Ordinarily, science probes aggressively but waits patiently. Partial evidence may be the stimulus for accelerated investigation. The literature abounds in scientific conjectures founded on limited studies and inconclusive data. Some of these conjectures are eventually falsified, some remain dormant and fail to mobilize research interest from other investigators, and others serve as catalysts for focused research activity.

When a hypothesis describes a connection between a human activity and a public health effect of some consequence, it will draw concern from certain stakeholders and interest groups who seek a speedy resolution. Thus, a political context emerges that adds a transscientific dimension to the disposition of the hypothesis. Although the pace at which nature's secrets are unlocked will not increase solely because the knowledge obtained may reduce human suffering, protect the environment, or resolve a controversy, public concerns may nevertheless influence the social process of discovery. For example, additional resources may be allocated for accelerated data gathering. Certainly, the potential catastrophic consequences of ozone depletion brought to bear widespread public pressure for resolution of the conjectural status of that hypothesis. Political pressures are inevitably reflected in actions taken by the scientific community. Various scientific constituencies begin to mobilize in support of reducing certain risks, even based on the limited evidence at hand. From a policy standpoint, it may be considered rational to act as if the hypothesis were true while retaining one's scientific skepticism. Taking action in the present to diminish or eliminate a human activity that is weakly conjectured to have catastrophic effects is ostensibly an insurance policy against the possibility of future disaster.

However, from an economic standpoint, it would be foolhardy to act on every hypothesis that posits an adverse outcome. First, the cost could be prohibitive. Some assurance is needed that there is a nontrivial probability of a significant effect. Second, competing hypotheses may confuse the plan of action and create social chaos. For example, there is the well-received hypothesis that chemical mutagens are likely to be human carcinogens. It has also been reported in the scientific literature that ingredients in such foods as mustard, peanut butter, herb teas, and beer are mutagenic and therefore may also be carcinogenic.[10] Policymakers acting on this hypothesis, who fail to consider alternative explanations, could put in place draconian regulations that would erode the confidence of the public in the integrity of science. Any scientific hypothesis is embedded within a wider system of beliefs. Similarly, an action principle in public policy must consider multiple factors, such as the nature of the consequences if the hypothesis proves true and no action is taken, the cost and effectiveness of strategies to mitigate the causal agent(s), and the nature of the consequences if action is taken and the hypothesis proves false.

The term *risk selection* refers to the social processes that elevate a risk hypothesis to the public agenda.[11] Rarely, if ever, is that elevation accomplished exclusively by the accretion of scientific knowledge. In 1948 the pesticide dichloro-diphenyl trichloroethane (DDT) was cited in a widely publicized book as a potential hazard to the environment, 14 years before Rachel Carson completed *Silent Spring*.* [12] Yet it took a talented nature

*Fairfield Osborn wrote *Our Plundered Planet* when he was president of the New York Zoological Society. The book had jacket reviews from Aldous Huxley, Robert Maynard Hutchins, and Eleanor Roosevelt and was published by Little, Brown and Co. (Boston). The author warned prophetically: "More recently a powerful chemical known as D.D.T. seems the cure-all. Some of the initial experiments with this insect killer have been withering to bird life as a result of birds eating the insects that have been impregnated with the chemical. The careless use of D.D.T. can also result in destroying fish, frogs, and toads, all of which live on insects. This new chemical is deadly to many kinds of insects—no doubt of that. But what of the ultimate and net result to the life scheme of the earth?"

writer[13]—and the serialization of her book in the *New Yorker*—to bring the issue of DDT into the light of public inquiry and another decade of debate before its use was banned in the United States.

Scientific concern over radioactive fallout from nuclear weapons testing brought little government action until strontium-90 was detected in humans. Similarly, the discovery of the Antarctic ozone hole in 1985 and the subsequent media response boosted public support for the Montreal Protocol on chlorofluorocarbon production. Dramatic scientific discoveries can provide the selective pressure that captures public imagination and transforms popular opinion into a powerful political force. These and other examples help us begin to understand the public response to a risk hypothesis. Sometimes it is a significant media event—usually one centering around human catastrophe (Bhopal, Love Canal, thalidomide)—that sets Congress and the regulators in motion. In other cases (ethylene dibromide, lead, dioxins, asbestos), no single event or discovery explains why federal actions are finally taken. It may be the sheer weight of evidence, litigation, media perseverance, or the dedicated work of a small group of unyielding advocates.

The Wingspread meeting of 1991 was the event closest to an official launch for the environmental endocrine hypothesis. Ironically, the meeting by itself hardly made a mark on society or the scientific community. Concern over endocrine disrupters registered nowhere near the emotional caliber of reaction to thalidomide, the discovery of the Antarctic ozone hole, or Three Mile Island in capturing the public's interest, fostering indignation, and solidifying scientific support.

In the case of endocrine disrupters, the general hypothesis is rather complex. Its fate is not linked to a single discovery. The general hypothesis is composed of a group of nested conjectures and theoretical scenarios of human and environmental effects based on a solid foundation of wildlife and laboratory studies. However, the mere accumulation of scientific studies is hardly sufficient to create a public issue. The process requires individuals who exercise a leadership role in synthesizing the extant literature and giving it a shape and a purpose. In this respect, Theo Colborn played a role analogous to that of Rachel Carson in bringing light and coherence to fragments of previously disparate scientific studies. But the conceptualization of a seminal public health hypothesis and its emerging visibility in the public consciousness are merely the first step toward preventing disease or transforming the way in which chemicals are regulated. Between research and action lie many formidable obstacles—not the least of which is the lack of convergence of the research studies and the challenge of meeting the scientific burden of proof.

Knowledge and Complexity

Most people would not question the proposition that more knowledge is better than less knowledge. There are, of course, exceptions: consider those who refuse to undergo genetic screening to determine whether they are carrying the genes for Huntington's disease or breast cancer. In cases in which one assesses the benefits of knowing to be minimal and the costs considerable, the individual at risk may choose ignorance over knowledge. For example, there are practical considerations that may prompt one to refuse genetic testing that could reveal vital health information. The prospect of losing one's health insurance as a result of the disclosure of a preexisting genetic abnormality may make *ignorance* preferable to *knowledge* in this circumstance.

In the social regulation of chemicals, more knowledge about chemical action on the biology of living things is always seen as an advantage by the regulatory bodies responsible for protecting public well-being and the environment. Yet two streams of thought have questioned the premise that more knowledge inevitably aids society in exerting control over the products of technology. The first argues, quite paradoxically, that the more we know, the less we know; the second maintains that certain claims to knowledge or partial evidence are dangerous because they are misleading. These claims have been labeled "junk science."

In biological investigations of the etiology of disease, scientists must resort to simplified models of the human biological system. For ethical reasons, humans cannot be used purely instrumentally as a means to discover truths about illness or well-being. Human subjects must be treated in accordance with the principles of the Nuremberg Code and federal guidelines on human subjects (as implemented by local institutional review boards) that protect the individual's right to informed consent and that give primacy to the health and best interests of the subject.

The methods used to gain causal or mechanistic knowledge of human disease must, because of these limitations, remain indirect and inferential. It is quite common that early evidence points in one direction, but that as more investigations are carried out, the system under study begins to look more complex. What appeared at first to be a straightforward causal relationship or strong association is subsequently found to be fraught with ambiguity. For illustrative purposes, let us consider the association of organochlorines and breast cancer in the context of the environmental endocrine hypothesis.

There is certainly good reason to posit a connection between organochlorine pesticides and breast cancer. Pesticides such as DDT metabolize into chemicals that accumulate in fatty tissue, and for this reason the breast is an excellent reservoir for them. There is a strong link between higher lifetime exposure to endogenous estrogen and breast cancer. Risk factors include early menarche, late menopause, and delayed child rearing. Age-adjusted breast cancer rates in the United States have shown a steady rise since 1940. During that period, there was a significant increase in the use of synthetic organochlorines in agriculture and industry. Studies performed in the 1960s indicated that organochlorines could induce mammary tumors in laboratory animals; however, those results did not prompt aggressive research into human health implications.

The discovery in the late 1980s that industrial and agricultural chemicals can mimic estrogenic activity made the connection between synthetic chemicals and breast cancer seem quite plausible and inspired new activism for pursuing the links through epidemiological studies of high-risk groups and communities with elevated cancer rates.Indeed, the early scientific studies seemed to be moving closer to unraveling the puzzle. In 1984 scientists reported higher levels of organochlorine residues in the breast tissue of breast cancer patients than in those without the disease. A similar result was found from studies of biopsy material from newly diagnosed patients undergoing breast surgery.[14] Two influential studies—one involving 20 cases and 20 controls[15] and the other 58 cases and 171 controls[16]—examined residues of 1,1-dichloro-2,2-bis ethylene (DDE, a metabolite of DDT) in women with breast cancer and controls who had not contracted the disease. They found a statistically significant association between DDE concentrations and the incidence of breast cancer. The language of the investigators reveals their cautious optimism about the evidence of an association, but the method they use precludes them from drawing any causal connections: "Our observations provide important new evidence relating low-level environmental contamination with organochlorine residues to the risk of breast cancer in women."[17] These studies reinforced the belief of many breast cancer activists that organochlorines in the environment are a primary cause of breast cancer.

But when a similar but more rigorous study was expanded to 300 women (150 cases and 150 controls) and involved collecting test serum 14 years before diagnosis (an advance over prior studies, which had collected fat or serum shortly before or after diagnosis), investigators found no association between DDE and breast cancer.[18] One might think the larger and more systematic study would take precedence and so resolve the issue.

However, a reanalysis of the data by ethnicity revealed that when Asian women were removed from the sample, an association between residues of DDE and breast cancer could be established.[19] The reanalysis results were consistent with other studies comparing American and Asian rates of breast cancer. Asian women in their native countries have lower rates of breast cancer than women of similar age in the United States. It has been conjectured that Asian diets, which are high in soya products, may neutralize the effects of exposure to high levels of organochlorines.

Although DDT has been classified as "possibly carcinogenic" to humans by the International Agency for Research on Cancer,[20] some scientists have argued persuasively that if it does play a role in breast cancer, it would be impossible to detect by epidemiological studies.[21] They maintain that the reason DDT might be linked to breast cancer is that it is estrogenic, although weakly so in animals. But the effect of DDT is complicated by the fact that one of its metabolites, DDE, was also found, under certain conditions, to accelerate the degradation of estrogen in some animals. Soto describes the scientific complexity of the issue: "Nature is not designed by an engineer. . . . We cannot get a clean classification for everything. For example, DDT is neurotoxic and it is also an endocrine disrupter with estrogenic effects. Yet its metabolites, or breakdown products, are antiestrogenic and have immunosuppressive effects."[22]

Thus far DDT has not been shown to have any appreciable estrogenic effects on women. Since so many women take oral estrogen for birth control and estrogen replacement therapy, the added potential effect of DDT would be masked and undetectable. This supposition is supported by the failure of occupational studies to show a strong link between organochlorines and breast cancer or endometrial cancer through endocrine mechanisms.[23] If there was a likelihood of detecting a link between DDT and breast cancer, one would think the prospects would be best among agricultural workers such as those in Mexico who spray DDT to control malaria—yet there has not been evidence of this connection to date.[24] Circumstantial evidence suggests there might be a link between intensive pesticide use in Hawaii—including that of many endocrine-disrupting pesticides, such as 1,2-dibromo-3-chloropropane (DBCP), DDT, kepone, and dieldrin—and the rising breast cancer rate. Researchers found correlations between higher pesticide exposure and the rate of breast cancer and abnormal cell growth among villagers, but the circumstantial evidence still falls far short of demonstrating a cause.[25]

Scientists have placed their hopes on statistically more powerful studies to resolve the conflicting evidence. Late in 1997 the results of the largest study to date testing the association between two organochlorines and the incidence of breast cancer was reported in the *New England Journal of Medicine.* A team of scientists investigated whether women with breast cancer had higher levels of either polychlorinated biphenyls (PCBs) or DDT in their blood plasma compared with a control group. The study and control samples were drawn from participants in the Nurses Health Study. Originating in 1976, the study focused on 121,700 married registered nurses from 11 states who had agreed to fill out a questionnaire on their health status and lifestyles every two years. Between 1989 and 1990 over 32,000 nurses sent in blood samples. The experimental group for the organochlorine study consisted of 240 women who were diagnosed with breast cancer after they had sent in their blood specimens. Each cancer subject was matched to a control subject who had not reported a diagnosis of cancer. The authors concluded that their data "[did] not support the hypothesis that exposure to DDT and PCBs increases the risk of breast cancer."[26]

The editors of the *New England Journal of Medicine* gave Stephen Safe the journal's prestigious editorial page to put closure to the suspicion that organochlorines were a cause of the rise in breast cancer rates. Safe had no difficulty generalizing from the study's affirmation of the null hypothesis:*

> Although the degree of exposure to estrogens over a lifetime is known to be a risk factor for breast cancer, the biological plausibility of the xenoestrogen hypothesis can be criticized on several counts. Most of the organochlorine pollutants, including PCBs and DDT or DDE, are only weakly estrogenic, and these compounds can both exacerbate and protect against mammary cancer in laboratory animals. . . . Moreover, the incidence of breast cancer has increased in industrialized countries over the past 20 years, but the environmental levels of most organochlorine contaminants have decreased as a consequence of strict regulations regarding their use and disposal . . . weakly estrogenic organochlorine compounds such as PCBs, DDT, and DDE are not a cause of breast cancer.[27]

But just as some scientists were putting to rest the connection between organochlorine pesticides and breast cancer, in December 1998 researchers at the Copenhagen Centre for Prospective Population Studies published

*The *null hypothesis* is the term statisticians use to assert that the independent variable (in this case organochlorines) makes no difference to the dependent variable (in this case breast cancer rates).

a study in *The Lancet* stating that women who had the highest levels of the pesticide dieldrin in their blood were at twice the risk of getting breast cancer. The researchers used blood samples taken in 1976 from more than 7,000 women who had participated in a heart study. They analyzed the blood from 268 of these women who developed breast cancer by 1993 and compared them with a cohort of 477 women in the same study who were free of the disease. Dieldrin has been shown in laboratory tests to attach to the estrogen receptor of breast cancer cells.[28]

Thus, with respect to breast cancer and organochlorines, as scientists probed more deeply and became more optimistic about promising associations, they were faced with mixed evidence that significantly weakened their conjectures. The scientific pursuit of the etiology of disease is a nonlinear accretion of information that may reveal ever deeper levels of complexity, in which evidence comes cyclically: the absence of support for a hypothesis is often followed by limited evidence confirming the hypothesis, then by new studies that do not show consistent patterns of association. Consensus may be reached when the studies begin to converge or when a definitive study (or some cluster of studies) yields incontrovertible evidence supporting the link between chemical exposure and disease. However, regulatory agencies use the initial divergence of studies as reason to delay or take no action.

In their book *Toxic Deception*, Dan Fagin and Marianne Lavelle note the liability of pursuing ever more fine-grained studies of chemical effects: "In toxicology, the temptation to perform study after study is particularly strong because the most important question—what are the health risks to humans?—is one that simply cannot be answered directly, because experimentation on humans is unthinkable. The endless pursuit of scientific knowledge can be dangerous in a regulatory system in which toxic chemicals are deemed innocent until proven guilty."[29]

Often, industry-funded studies focus on pinpointing countervailing evidence to a public health hypothesis. The more detailed and fine-grained the studies, the more opportunities for ambiguity, regulatory error, and controversy that work against diminishing chemical use or finding safer substitutes. In Carl Cranor's superb work *Regulating Toxic Substances*, he argues persuasively that "the attempt to have more fine-grained knowledge about the risks from a particular substance may actually increase the number of mistakes agencies make and may add to the misconceptions associated with risk assessment."[30] By the 1990s the chemical industry had launched an ideological offensive against the use of scientific hypotheses implicating chemicals in human disease. Couched in terms of good and bad science, this campaign has won the support of the mainstream media. So whereas some stakeholders believe that partial knowledge justifies regulatory action, and that such action should not be stymied by the fact that all the details are not consistent, other stakeholders consider policy actions faulty without a scientific analysis that resolves the uncertainty and inconsistency in the details.

Junk Science, Sound Science, and Honest Science

There was a time when scientists of all political stripes rallied against creationism, the quintessential example of religion masquerading as science. They also protected their disciplines from the pseudoscientific claims of extraterrestrials and life after death that so irked the scientific establishment. Although many purveyors of metaphysical beliefs wish to gain the imprimatur of science, there have also been cases in which certified scientists, respected in their fields, were the source of unsubstantiated scientific theories. For a short while scientists took seriously the claims made by a chemist that fusion of atoms to produce energy could be achieved through a relatively simple chemical process called "cold fusion." When scientists could not replicate experiments that purported to demonstrate the effect, they began to take a hard look at the peer review process to ascertain how preliminary publications describing unverifiable results could have been published. In some respects, the cold fusion episode was a success story. The claims of the cold fusion advocates were put to the test and failed to meet the minimum standards of evidence required to validate a scientific hypothesis.

In the 1970s, the passage of new environmental laws, and the explosion of public awareness that the products of so-called industrial progress can be harmful to human health, were followed by escalating litigation, particularly class action suits over drugs and toxic substances such as diethylstilbestrol (DES), asbestos fibers, and hazardous waste sites. The courts were overwhelmed with tort cases that pitted expert against expert, résumé against résumé, in an effort to persuade juries. The rules of evidence were strained. Judges were unable to distinguish between the quality of the experts and the conflicting scientific analyses they presented to the court. Jurors had a difficult time evaluating dueling experts. Many were academics who either supplemented their modest incomes by appearing as highly paid expert witnesses for plaintiffs or were hired by companies to defend them against liability claims.

The issue of conflicting expert testimony in the courts first arose in the 1920s over the use of evidence from a lie detector device. In *Frye v. United States* (1923), the United States Court of Appeals for the District of Columbia Circuit ruled against the admissibility of evidence from a systolic blood pressure deception test, the precursor to the modern polygraph test. The appeals court issued a statement of principle that, to be admissible in the courts, scientific evidence must be sufficiently established to have gained general acceptance in its particular field. This principle, which became known as the Frye rule, had the effect of reducing judgments against companies since it limited the use of speculative theories of causation in trials. In 1993 the Frye rule was superseded by a Supreme Court decision in the case of *Daubert v. Merrell Dow Pharmaceuticals* that gave the trial judge the responsibility for determining the admissibility of technical evidence by applying certain criteria to the expert testimony. Among the standards to be applied were these: Has the scientific knowledge been tested? Was it submitted to peer review and publication? Is the theory or technique generally accepted within the scientific community?

Growing out of the litigious science involving environmental disputes, a new term, *junk science,* was introduced in the 1990s to discredit certain claims made during expert testimony. The introduction of the term had a divisive effect within the scientific community. No longer was the issue pseudoscience versus authentic science, but rather that science which threatens corporate profits and economic growth versus the rest of science. Politically conservative scientists use the expression to cast aspersions on the honest research and hypothesis-framing of many credible scientists in government and universities. Indeed the one common characteristic of all the claims cited as junk science is that such claims, if acted upon, would cost certain industrial sectors substantial sums in tort cases, product liability judgments, or product and process substitutions.

The environmental endocrine hypothesis is only the latest of many reputable scientific hypotheses involving environmental health effects that have been accorded entry into the pantheon of theories and hypotheses labeled junk science. Other public health risk factors in this group include DES, asbestos, radon, the drug Benedictin, the Dalkon Shield, global warming, silicone breast implants, Alar, and electromagnetic fields.

The metaphor of junk science was popularized in a book by Peter Huber titled *Galileo's Revenge: Junk Science in the Courtroom.*[31] Huber uses the term *junk science* to disparage the testimony of some scientists who serve as expert witnesses. The concept was elevated to prime-time status in a series of four television reports aired in 1997 on *Good Morning America* and an hour-long ABC News special, all hosted by John Stossel and titled "Junk Science: What You Know That May Not Be So." Stossel featured several scientific hypotheses including the role of vitamin C in cutting back on colds, the role of salt in causing high blood pressure, the risks of breast implants, the human health hazards of dioxin, and the emergence of a new syndrome called multiple chemical sensitivity.

In their book *Betrayal of Science and Reason,* Paul and Anne Ehrlich describe Stossel's reception at the 1995 annual national conference of the Society of Environmental Journalists: "John Stossel of ABC's *20/20* was pressed by a reporter about whether he still considered himself a journalist in view of the tens of thousands of dollars he received in speaking fees from chemical companies and other business groups. Stossel replied, 'Industry likes to hire me because they like what I have to say.' He then added that he supposed he was no longer a journalist in the traditional sense but rather a reporter with a perspective."[32]

The best that one can say about the purveyors of the term *junk science* is that they force us to think about the different confidence levels associated with scientific hypotheses, they demand that we not ignore negative evidence, and they implore us to improve the peer review process. The fact that a hypothesis is possible does not make it probable or true.

The worst that we can say about the junk science initiative is that it smacks of an ideology that opposes investigations into consumer products and environmental hazards. The very label trivializes the process of hypothesis generation, evidence gathering, and adjustment. Hypotheses have been labeled junk science because the sample from which they are derived is small, or because there is a piece of countervailing evidence, or because they are unable to explain all they purport to. Certainly these are all sound reasons to question any explanation and to reexamine it in the light of alternative hypotheses. They may even be a reason to question policy enacted based on the hypothesis.

But the term *junk science* implies that the studies advanced in support of certain hypotheses are methodologically deficient, namely, that they do not adhere to the canons of scientific integrity or the quality assurance standards necessary to assert a causal connection. Many superbly executed studies are guided by hypotheses that are eventually proven false. Those who cry "junk science" miss the distinction between process and outcome. On many an occasion even good science guided by a strong and reliable hypothesis will, on further investigation, result in the falsification of that hypothesis.

The concept of "sound science" derives from criteria within the scientific community. The canons of a scientific discipline are embodied in its peer review process. Most scientists are sufficiently well trained to distinguish among pure conjectures, tentative or partially confirmed hypotheses, and strongly supported causal explanations. Scientists are cautious about inferring causal relationships from data when there are no controlled experiments, when the reproducibility of results is in doubt, when the range of evidentiary support is narrow, or when the empirical evidence is exclusively associational (the putative cause is found in the presence of the effect). Popular renderings of pure or highly speculative conjectures blur the epistemological distinctions that are basic to scientific literacy.

The environmental endocrine hypothesis represents a scaffolding of propositions, some tested and some untested, built on an extensive and growing body of field studies, controlled animal experiments, epidemiological data, and in vitro studies drawn from diverse disciplines. Even the strongest proponents of the hypothesis are cautious about making causal claims that link exposure to endocrine disrupters (as determined by animal or in vitro assays) to human health effects. The so-called Weybridge Report on endocrine disrupters issued by the International Organization for Economic Cooperation and Development stated that "there is insufficient evidence to definitively establish a causal link between the health effects seen in humans and exposure to chemicals."[33] But such caveats are not grounds for labeling the general hypothesis junk science. Support for the hypothesis rests on laboratory and wildlife studies and some occupational exposure evidence.[34] No credible scientist has disputed the findings linking chemicals to endocrine effects in a variety of animal systems. As *The Lancet* reported in 1996, "The potential of environmental estrogens to exert deleterious effects on wildlife is now well established. The effects of organochlorine compounds in localized populations at risk have also been identified in some instances."[35]

There is also divergence among the proponents of the endocrine hypothesis. Some believe that endocrine disrupters are the cause of rising breast cancer rates. Others are highly skeptical about the breast cancer link but believe that sperm count decline is a more likely outcome of fetal exposure to chemicals. Some are persuaded by animal models and human epidemiological data that support the sperm decline hypothesis. For example, rats and hamsters exposed to certain endocrine disrupters prior to and shortly after birth have reduced sperm counts when they become adults.

Critics of the endocrine hypothesis point to many holes in the lattice of supporting evidence. Gaps in the data or conflicting evidence challenge proponents to modify or sharpen the general hypothesis and explain the anomalies. Notwithstanding the gaps, the general theory of endocrine disrupters has established that chemicals can adversely affect the endocrine system of animals at certain exposures and during certain developmental windows. The six scientific consensus statements (Appendix B) are testimony to this. The controversy over endocrine disrupters is not over such basic principles as the ability of some synthetic chemicals to bind to animal and human hormone receptors and initiate gene expression. The debate is over the invocation of these principles to predict or explain human health effects. Because of the new emphasis on the endocrine effects of chemicals, biomedical researchers are increasingly setting their sights on and discovering new biological actions of synthetic chemicals.[36]

It was the threat to industrial interests posed by a hypothesis that linked chemicals to disease that precipitated the "junk science" attacks on the hypothesis. The concept of junk science, which has its origins in the juridical uses of adversarial experts in tort litigation, has been misused to discredit sound hypotheses posed in the course of scientific inquiry before they have been fully tested. The purveyors of the concept demand a high burden of proof before implicating a chemical and a much tighter control of the scientific data required to assess risk. They advocate leaving to the most conservative institutions of science the role of informing policymakers. The former chief executive officer of Olin Chemical Corporation wrote in an editorial published in *Chemical and Engineering News* titled "Combating Junk Science": "I would propose that we strengthen the vital role that NAS [the National Academy of Sciences] and the science advisory boards and committees play in advising EPA [the Environmental Protection Agency], the Food and Drug Administration, and other agencies."[37] The implication of this proposal is that only selected science advisory boards should be invested with the authority to screen risk hypotheses.

The current euphemism for reducing chemical regulation is "science-based policy." In essence this approach boils down to: don't regulate until you have complete confidence in the causal connections. Cranor describes the approach:

> One reaction to present agency practices is to recommend that risk assessment procedures should be based on better science, and that

> they should be more accurate, producing predictions of more "real" risks and fewer "theoretical risks." . . . This approach advocates using only the most complete and accurate science in order to arrive at estimates of risks of harm to human beings from toxic substances. On this view, if peer-reviewed scientific information does not provide a *complete* understanding of the disease-causing mechanisms for a careful evaluation so that an accurate evaluation of the risks is possible, there should be no regulation.[38]

Instead of focusing on a constructed ideological divide between "sound science" and "junk science," I propose we introduce the category "honest science." How do we know that the scientists who are advising the federal agencies are sufficiently disinterested in the commercial interests at stake that they will render a decision in the public interest? Honest science is science that discloses financial interests and other social biases that may diminish the appearance of objectivity in the work. In litigation, expert witnesses are often cross-examined about their financial interests in the case. Yet the financial interests of scientists serving on advisory boards to government agencies and panels formed by scientific associations are rarely revealed to the public. In one study of the membership of the NAS, my colleagues and I reported that in 1988 at least 39 percent of biological scientists had undisclosed ties to companies. The NAS conducts many studies for government agencies by convening expert panels but rarely releases to the public the conflict-of-interest disclosures filed by its panel members. A 1996 study of 14 biomedical journals revealed that 34 percent of the articles published in 1992 by Massachusetts-based contributors had at least one lead author with an undisclosed financial interest in the research being reported.[39] None of the personal financial interests of the journal's authors was disclosed in the nearly 800 publications surveyed.

Thus, by making the selection of those risk hypotheses deserving of public response the purview of certain expert panels, without revealing the potential conflicts of interest of the scientists who serve on those panels, a bias toward private interests over public interests may be built into the process. Full disclosure does not guarantee that there will not be bias or conflicts of interest, but it does raise the stakes in favor of scientific neutrality and objectivity.[40]

How should society respond when evidence that supports a specific chemically induced endocrine effect is undefinitive but accumulating, suggestive, or circumstantial? To investigate this question further, let us examine the normative assumptions that connect scientific uncertainty with policy responses.

Scientific Uncertainty and Social Action

What is the legal or ethical basis for action when some segment of the scientific community advances a hypothesis that asserts a public health risk based on limited and inconclusive evidence? This is not an issue of a pseudoscientific claim masquerading as science, but rather an issue of policymaking under conditions of scientific uncertainty. In the case of chemical hazards it means we have limited direct scientific evidence, only indirect or circumstantial evidence, that a chemical is guilty. In science a hypothesis can remain in limbo for years or even decades before there is a resolution, that is, it is either falsified or confirmed. In some cases, years of failed efforts either to confirm or to falsify a hypothesis may result in its abandonment by scientists, particularly when alternative explanations are available.

However, policymakers do not have the option of waiting it out. They can decide *not to act* on the hypothesis, thereby accepting, on behalf of society, a potential (albeit perhaps remote or hypothetical) risk, or they can decide to act on the basis of limited knowledge. But inaction is a form of action, and it can have public health or environmental consequences.

The idea of acting on limited evidence, rather than waiting for definitive causal evidence, was gaining support from the international community during the second North Sea Conference, held in London in November 1987. The goal of the conference was to seek ways to protect the marine ecosystem of the North Sea from bioaccumulating persistent organic pollutants. The consensus document from that conference, known as the London Declaration, referred to a principle of precautionary action subsequently known as the "precautionary principle":

> In order to protect the North Sea from possibly damaging effects of the most dangerous substances, a precautionary approach is necessary which may require action to control inputs of substances even before a causal link has been established by absolutely clear scientific evidence. . . . [The parties] therefore agree to . . . accept the principle of safeguarding the marine ecosystem of the North Sea by reducing polluting emissions of substances that are persistent, toxic, and liable to bioaccumulate at [the] source by the use of the best available technology and other appropriate measures. This applies espe-

> cially where there is reason to assume that certain damage or harmful effects on the living resources of the sea are likely to be caused by such substances, even when there is no scientific evidence to prove a causal link between emissions and effects ("the principle of precautionary action)."[41]

The theme of public responsibility in the face of limited scientific evidence was also discussed during the 1992 Earth Summit in Rio de Janeiro (officially referred to as the United Nations Conference on Environment and Development). It was particularly germane to the complex hypothesis of global warming. One of the regional preparatory conferences for the Rio meeting was held in Bergen, Norway, in May 1990; it was attended by the environment ministers of 34 countries. The Bergen Declaration that was issued at the conclusion of the conference affirmed the precautionary principle, calling for some potentially dangerous activities to be restricted or prohibited "even if there is no conclusive scientific proof linking the particular substance or activity to environmental damage."[42] Some interpreters of the Bergen Declaration restrict the scope of the principle to those activities that threaten to produce serious and irreversible harm.* [43] Others offer a more liberal interpretation and advocate a "better safe than sorry" approach to a wide spectrum of environmental conditions, regardless of the gravity of the consequences.

A number of European nations have integrated, at least in spirit, versions of the London Declaration into their domestic laws and regulatory behavior. In Germany, for example, the concept of a *Vorsorgeprinzip* (literally translated as precautionary principle) implies that "environmental protection policy should be preventative instead of reactive, employing avoidance and reduction of emissions technology at the sources."[44]

Considerable debate has focused on the threshold conditions under which the precautionary principle should be brought into play. The threshold problem is a major consideration cited by critics for opposing the principle. They argue that there may be a wide gradation of evidentiary support in published scientific articles, ranging from purely theoretical risks to fully justified causal knowledge. It is within this range that the politics of the precautionary principle is most pronounced.

*According to Perrings, "The principle can be interpreted as saying that if it is known that an action may cause profound and irreversible environmental damage which permanently reduces the welfare of future generations, but the probability of such damage is not known, then it is inequitable to act as if the probability is known."

Opponents of the precautionary principle claim that

— It justifies regulation on the thinnest veneer of hypothetical evidence. Even the purest speculation can justify banning a chemical. Chemicals are guilty until proven innocent.
— It fails to consider the relative risks of alternative chemicals.
— It neglects the cost-benefit calculations involved in regulating or banning a chemical substance.

In a public debate with the authors of *Our Stolen Future,* Elizabeth Whelan of the chemical industry–oriented American Council on Science and Health cast the book in the same negative light as the principle it advocates:

> The precautionary principle seems to dominate the book of my fellow panelists. Typically, it is invoked in situations where the scientific evidence is extremely tentative but the potential for arousing fear is great. . . . I don't buy the precautionary principle for several reasons. First, it always assumes worst-case scenarios. Second, it distracts consumers and policymakers alike from the known and proven threats to human health. And third, it assumes no health detriment from the proposed regulations and restrictions. . . . It overlooks the possibility that real public health risks can be associated with eliminating minuscule hypothetical risks.[45]

The precautionary principle affirms that one does not have to wait until definitive evidence of a health risk has been demonstrated before regulatory action can be taken. The principle implies that policy can precede science. It is thus at odds with the favored industry euphemism "science-based policy"—meaning not only policy grounded in sound science but also policy whose enactment awaits the publication of conclusive evidence. As an illustration, a representative of a chemical company responding to Theo Colborn's recommendations to regulate endocrine disrupters was quoted as saying: "Just because something is shown to be an endocrine disrupter in a laboratory test doesn't mean that in the real world it's going to be a problem . . . there may not be any exposure. You have to relate it to actual exposure in the environment before you can say it is a risk."[46]

The debate over whether science should precede policy or vice versa is rooted in two concepts: the *burden of proof* behind public decisions and *errors in decisionmaking.* The concept of burden of proof is the wedge that separates two divergent social beliefs about chemicals and technology. Some skeptics argue that chemicals should be proven safe before they are accepted into industrial use and consumer products; others are satisfied with the general strategy that chemicals are considered safe until proven unsafe. We are faced with a dilemma. On the one hand, we cannot *prove* that something is safe; we can only falsify hypotheses that assert a hazard. On the other hand, it is extremely difficult to demonstrate, with a high level of scientific confidence, that chemical exposures at low levels are hazardous to humans. In theory it should be easier to prove the negative—"chemical X is unsafe"—than the positive—"chemical X is safe." In practice, however, the standards for demonstrating that low-level exposure to certain chemicals is unsafe have been difficult to meet, and based on the number of chemicals actually banned from industrial use, they have in fact only rarely been met.

Given these circumstances, one might argue that a high burden of proof means that risk hypotheses pertaining to chemical exposures ought to be tested under multiple conditions and that to be considered "safe" a chemical has to pass this battery of tests. We might call this the hurdle theory of risk management. In this view, chemicals are considered guilty, that is, hazardous, until they meet the standard for innocence. A still higher burden of proof would place any synthetic chemical on a forbidden list unless all of the following could be demonstrated: there are no natural substitutes, the chemical has unequivocal social benefits, and the chemical has passed the battery of tests allowing it to meet the standard for innocence. These burdens of proof have economic and social ramifications that inevitably will be debated by economists and challenged by the industries whose very existence depends on the continued production of synthetic chemicals.

The concept of burden of proof applied to the regulation of suspect chemicals has been illuminated by the classification and analysis of two types of errors discussed in basic statistics.[47] In testing a hypothesis that posits a causal connection between a chemical and an effect, a Type I error is committed when the prevailing evidence leads us to accept the causal connection, when in reality it fails to exist. This is the error of overzealous risk identification. Suppose our hypothesis is the following: Endocrine-disrupting chemical A is harmful to humans. The null hypothesis posits no effect: Chemical A shows no adverse effects on humans. A false positive or Type I error rejects the null hypothesis and asserts an effect.

A Type II error occurs when, on the basis of prevailing evidence, we reject a causal connection between a chemical and an effect, when the effect actually exists (when the hypothesis linking the chemical to human disease is true or, alternatively, the null hypothesis is false). A false neg-

BELIEF IN HYPOTHESIS based on prevailing evidence		**TRUTH OF HYPOTHESIS** Null hypothesis is true: Chemical A is not associated with human disease. H is false	**H = Chemical A is harmful to humans.** Null hypothesis is false: Chemical A is associated with human disease. H is true
	F	No error	TYPE II ERROR False negative Industry choice Effects are denied when there are effects
	T	TYPE I ERROR False positive Consumer choice Effects are affirmed when there are not effects	No error

Figure 1

Type I and Type II errors under conditions of uncertainty

Source: Adapted from Cranor (1993:15)

ative (Type II error) accepts the null hypothesis of no effect and is consistent with underzealous risk identification and with the implication that the conditions of chemical hazard remain undetected and unregulated (see Figure 1).

Shrader-Frechette asks, given a situation of uncertainty, which error is more serious.[48] Within the framework of civil liberties law, the conventional response has been that it is a more serious error to convict an innocent person (for the hypothesis "P is guilty," a Type I error) than it is to free a guilty person (a Type II error). But what about tort or regulatory law? Suppose the hypothesis is "Chemical A is harmful to humans." Is it a more serious societal error to fail to act on chemical A when it is harmful than it is to regulate chemical A when it is not harmful? Representatives of the chemical industry consider a Type I error more serious than a Type II error since it reduces their profits. In contrast, consumers and public interest groups generally consider a Type II error more problematic than a Type I error since it underestimates the hazards posed to humans.

Shrader-Frechette correctly notes that minimizing false positives—assertions of a chemical effect where none exists—is a priority of science. Scientists prefer skepticism (false negative) over the premature adoption of a false hypothesis (false positive). Philosopher Abraham Kaplan notes that "the scientist usually attaches a greater loss to accepting a falsehood than to failing to acknowledge a truth."[49] However, for the public health community the reverse is often the case: false positives are preferable to false negatives. The public health perspective prefers to err in overstating risks rather than in understating them since the former involves greater economic costs while the latter involves greater morbidity and mortality.

Nicholas Ashford has argued that the dilemma represented by the choice between Type I and Type II errors may be a false one. By failing to regulate a chemical on the basis of suggestive evidence, Ashford notes that we might commit a Type III error—which he describes as failing to protect environmental health and failing to support a substitute technology that would result in a win-win condition in the economy.[50]

We may be inclined to think of the debate over Type I and Type II errors as one best relegated to academic venues, with no policy relevance. However, policy analysts have been hammering away at this issue for years, and the result has been a new awareness of the value assumptions implicit in choices designed to minimize errors. In its recommendations for the development of an endocrine disrupter screening and testing program, the EPA's Endocrine Disrupter Screening and Testing Advisory Committee (EDSTAC) stated that the screening strategy (the first tier in its proposed battery of tests) should have as its primary objective the minimization of false negatives or Type II errors while permitting some acceptable level of false positives or Type I errors. In its proposed tests on whole animals (second-tier tests), the goal of the committee was to minimize both Type I and Type II errors, as we shall see later in this chapter.

In summary, the precautionary principle is consistent with minimizing Type II errors for hypotheses that assert a link between chemical exposure and environmental or human hazards. Critics of the precautionary principle seek to discredit its application to a given hypothesis by tagging the hypothesis with the pejorative term *junk science* when the issue is not about science but about public policy, namely, how we are to handle uncertainty and risks and where the burden of proof should rest. Of course if there is no evidence backing a hypothesis that links a chemical to human disease, then we may speak of the hypothesis as purely speculative, even though it might be testable. Alternatively, if the science that is used to frame a hypothesis is faulty, then there are grounds for treating the hypothesis as discredited. There is no basis for claiming that the generalized environmental endocrine hypothesis is either purely speculative or based on faulty science. Within the broad outlines of the hypothesis there is considerable controversy over finer-grained subhypotheses, for example, whether the levels of human exposure to certain chemicals are high enough to effect a response. Therefore we must ask which regulatory path should we take: the aggressive or the conservative? Either minimize Type I errors and let the economy bloom, or minimize Type II errors and ensure that future generations do not curse us. Are we involved in a devil's gamble?

Limits of Current Regulatory Policy

Conjectures about the relationship between synthetic chemicals and the endocrine system in fetal development were raised as early as 1979, according to Stone.[51] In its broadest formulation, the environmental endocrine hypothesis serves as a unifying framework within which to consider a disparate class of reproductive and hormone-related pathologies in a variety of species. The messages about endocrine disrupters communicated to the public in the environmental literature and the press are provocative and profoundly disturbing. They include intersex characteristics (reproductive organs with combined male and female features) found in marine snails, fish, alligators, fish-eating birds, marine mammals, and bears; declines in human sperm count of as much as 50 percent; increased risk of breast cancer; small phalluses in Florida alligators resulting from pollution; penises found on female mammals; undeveloped testes in Florida panthers; masculinized female wildlife with a propensity to mate with normal females;[52] and cognitive deficiencies in children.

The public policy ramifications of even a small subset of these outcomes are significant. Each of the subhypotheses of the endocrine hypothesis is associated with an evolving body of evidentiary support. Thus, the general hypothesis is supported by a lattice of interconnecting but independent hypotheses, many describing associations rather than causal links between chemicals and effects. To date, despite the accumulating evidence, the environmental endocrine hypothesis cannot claim a single dramatic episode or discovery capable of turning public opinion into a potent force for political change or, for that matter, capable of inspiring a broad scientific consensus on the risks to human health.

Nevertheless, scientists point to effects that, in their view, represent a clear and present environmental danger. Increasingly, new scientific panels are voicing their beliefs that the evidence for some subhypotheses within the general hypothesis is compelling. Among the best-documented effects of endocrine disrupters is the dramatic reduction of the alligator population of Lake Apopka, Florida, the state's fourth largest body of fresh water. The lake was contaminated from years of pesticide runoff, effluent from a sewage treatment plant, and a pesticide spill. Alligator eggs collected from the lake were found to have concentrations of endocrine disrupters, such as DDE, 10,000 times greater than those normally found in the blood of newborn alligators.[53] In 1998, reproductive problems were also found in alligator populations in Florida lakes that had not experienced the intense contamination of Lake Apopka, indicating that alligator habitats may be threatened throughout the state. One possible explanation for the reproductive problems is that chemical runoff from farms may be leaching into alligator habitats. Another documented effect of endocrine disrupters is the disappearance of trout from the Great Lakes. The source of the problem has been traced to dioxin-like pollutants from industrial effluent.[54] Both examples point to inadequacies in the current system of environmental regulation. For some endocrine disrupters implicated in these cases, such as DDT and PCBs, regulatory actions have been taken. However, many other chemicals with endocrine-disrupting effects remain in wide use and, according to environmentalists, merit immediate regulatory attention.

Anticipating congressional action, U.S. regulatory agencies began to take notice of hormone disrupters. As we have seen, the EPA and the Department of Interior contracted with the National Academy of Sciences to undertake a study of endocrine disrupters.[55] The EPA issued grants to fund the study of linkages between ecological and human health impacts of xenobiotic chemicals. It also began collaborating with other agencies, such as the National Institutes of Health, in an effort to uncover the causal mechanism of estrogen-mimicking chemicals.

We have already seen some of the obstacles associated with elevating a risk hypothesis into the policy arena. Let us assume that in the not-too-distant future the conditions are such that, despite the current antiregulatory mood in the United States, a decision is made to regulate endocrine disrupters. To what extent are we prepared for this step? Are the existing laws sufficient to address the issues? Will we be able to add this responsibility to the already heavily burdened regulatory system?

Several general conclusions can be drawn at the outset about U.S. regulatory policy directed at hazardous chemicals. The policy has, for the most part, approached the management of hazardous substances chemical by chemical. Of the tens of thousands of chemicals that have been introduced into agriculture and industrial production, very few have been explicitly banned from use; for those that have, it has taken decades to remove them. A few chemical groups have either been banned or their use significantly restricted. For example, PCBs are a family of over 200 chemical variants (congeners) that were banned for all manufacturing processes in 1979, and the family of chlorofluorocarbons are being phased out of many products. However, these are the exceptions. Most regulated chemicals may be used within permissible limits.

Chemicals are regulated differently according to when they were introduced into commerce (some have been grandfathered in), their applications (food additives versus insecticides), their putative effects (cancer-causing versus neurotoxic effects), the populations that are exposed to them (children versus workers), and the pathways of human exposure (air, water, soil, prepared food, occupational).

There are six general ways in which governments attempt to control hazardous or potentially hazardous materials: they conduct research; they establish economic incentives for substitution or reduced use; they enact legislation; they issue regulations; they undertake public education; or they use moral persuasion (voluntary guidelines) to change consumption or production patterns. The most highly publicized federal action on an estrogenic chemical occurred when the EPA banned the use of DDT in 1972, primarily because of its impact on the reproduction of birds. At the time there was only circumstantial evidence from animal studies about DDT's possible carcinogenic effect on humans. The prohibition of DDT, by virtue of its public visibility through Rachel Carson's work, was rather untypical. Most pesticides that have been banned or severely restricted, such as chlordane or aldrin and dieldrin, also exhibited carcinogenic effects on mammals, but their regulation was accompanied by less national attention and fewer behind-the-scenes legal challenges. Cancer and acute toxicity have been the dominant end points guiding the regulation of pesticides. In 1990, a report by the General Accounting Office cited the insufficient attention given to reproductive toxicity by federal agencies. According to the report, at most 7 percent of the synthetic chemicals in use had been tested to determine whether they might harm the reproduction and development of animals.[56]

Two primary statutes structure the EPA's regulatory course of action for chemical substances: the Federal Insecticide, Fungicide, and Rodenticide Act (FIFRA), first passed in 1947 (and amended several times since then, including 1972, 1988, and 1996) and the Toxic Substances Control Act (TSCA), enacted in 1976. The older and stronger of the two acts, FIFRA is designed to regulate the use of pesticides. The 1972 amendments provided for premarket screening of pesticides. Manufacturers must demonstrate that a pesticide does not present "unreasonable adverse effects" to human health and the environment before being able to register it. The determination of "unreasonable adverse effects" must take account of the agricultural benefits afforded by the pesticide. Specifically, there are currently approximately 24,500 formulated pesticide products made from about 900 pesticide active ingredients and about 2,500 chemicals classified as "inert" ingredients. The figure for total formulated pesticide products is down from about 50,000 before the re-registration of pesticides was initiated under the 1988 revision of FIFRA. The EPA generally tests only active ingredients rather than formulated products unless special circumstances require the testing of a particular pesticide formulation. The 1996 Food Quality Protection Act gives the agency authority to add "estrogenic" to the list of undesirable effects in reviewing pesticides, but it provides no criteria for testing endocrine disrupters.

There are several thousand formulations of pesticides already in use that have not been tested under the criteria established for new pesticides. Under a congressional mandate, the EPA has been asked to reexamine all pesticides currently in use for their health and environmental impacts. Assessing the effects of chemicals on the endocrine system has not been an explicit component of the testing protocols for pesticides. However, it is within the authority of the EPA under FIFRA to broaden the safety criteria according to which pesticides are evaluated. Prior to adding new criteria to its pesticide data requirements, the EPA would ordinarily prepare a scientific background paper justifying the action, convene a scientific advisory panel, and call for public comment on the draft guidelines.

Within a year after Congress had passed the TSCA, government officials estimated that the candidate list for the mandatory inventory of chemicals in commerce was approximately 35,000. Under the act, the EPA was given the authority to ban or restrict the manufacture, processing, distribution, commercial use, or disposal of any chemical substance or mixture that presents an unreasonable risk to human health or the environment. Prior notification by the maker is required for the manufacture of any *new* chemical substance (the so-called premanufacturing notice or PMN) and for the manufacture or processing of an existing chemical substance for a significant new use. There are three main provisions of the act. The EPA may require *testing* of any chemical substance or mixture if it has reason to believe the chemical *may present* unreasonable risk by virtue of its inherent toxicity or as a result of significant exposure to it. The authority of the EPA to *restrict* a chemical requires a finding that it *will present* a risk. Finally, the EPA may require record keeping and reports on chemical use and manufacture.

The TSCA requirements are different for new and existing chemicals. A manufacturer is required to issue a PMN before it can introduce a new chemical into production. The law gives the agency 90 days to review any new chemical that is being considered for industrial use. The agency uses limited criteria for the assessment of the new chemical (for example, it frequently applies structure-activity analysis, which assumes that chemicals with similar structures will exhibit similar properties). The agency also maintains an inventory of chemicals currently in use. The EPA had listed approximately 62,000 chemicals on its original inventory in 1979. As of August 1998 there were 75,500 chemicals on this list, among which were 2,643 inorganics, 48,697 organics, and 24,160 polymers.[57] Many of these chemicals had been grandfathered into use and were therefore not subject to rigorous assessments. In the first half decade the TSCA was in effect, the EPA received about 1,500 PMNs for new chemicals being produced each year and required some form of testing in fewer than 10 percent of the cases.[58] Fewer than 1 percent of new chemicals were subject to complete health hazard assessment.[59]

In their review of the TSCA, Fagin and Lavelle noted that "when the Toxic Substances Control Act was passed in 1976, it seemed, on its face, to give the federal government enormous power to protect the public from dangerous products. But the law stipulated that the EPA must weigh the likely cost of its decision against the potential benefits and prove that it had chosen to use its power in 'the least burdensome' way to industry. Consequently, the EPA has been able to regulate only a relative handful of chemicals."[60] The TSCA was soon deemed outdated by both opponents and proponents of strong environmental regulations. In her polemic against environmentalists, Edith Efron quotes insiders in the regulatory world as saying that even with all the screening facilities mobilized in the United States, "there are simply not enough toxicologists, pathologists, animal suppliers and laboratory facilities to test all chemicals."[61]

A General Accounting Office study of the TSCA reported that an EPA committee had recommended a total of 386 substances for testing over a 14-year period and that the agency had complete test data for only 6 chemicals.[62] Although the agency has statutory powers under the TSCA to remove a chemical from use or limit its use, it must meet a significant burden of proof to exercise its authority to require screening for any end points. Thus, given the limited powers conferred by the TSCA, it is difficult to imagine that effective endocrine-function assessment criteria will be introduced, without some amendments to the law, when the more established end points of chemical assessment—namely acute toxicity and carcinogenicity—are themselves underutilized. Moreover, the TSCA is probably more responsive to cost-benefit considerations and proprietary constraints than either FIFRA or the food additive laws. As Portney notes, "The legislation [TSCA] has not been very effective, largely because there are no clear-cut testing procedures or standards to determine whether a chemical does indeed present an imminent hazard, and because many companies have claimed that their products are proprietary—that they and only they have a right to know what the chemicals are."[63]

With the prospect that a nontrivial percentage of the tens of thousands of chemicals in current use might be endocrine disrupters, the government is faced with a serious policy challenge. Some prioritization must be developed for screening chemicals currently in use. Under the current case-by-case approaches underlying much of chemical regulation, it would take decades and substantial sums of money to meet a new set of regulatory goals based on the potential hazards of endocrine disrupters. According to Hynes, "On paper both laws [TSCA and FIFRA] purport to solve the problems of preventing dangerous chemicals from being used commercially. At best, they keep only the worst new chemicals off the market."[64] If the testing of putative carcinogens is any example, definitive studies will be costly and time consuming. The preferred mammalian studies would have taken from three to five years to complete and cost upwards of $1 million per chemical in the mid-1980s, according to Lave and Upton.[65] One must anticipate multimillion-dollar price tags for such studies in current dollars.

Yang estimates that the United States has acquired good toxicological data on about 500 chemicals, less than 0.1 percent of the number he believes are currently in commerce—a total he places at more than half a million, eightfold higher than official EPA estimates:

> The U.S. National Toxicology Program (NTP) and its predecessor, the National Cancer Institute's Carcinogenesis Bioassay program, collectively form probably the world's largest toxicology program. In its near-25 yr of operation, about 500 chemicals have been studied for carcinogenicity and other chronic toxicities. These chronic toxicity/carcinogenicity studies and the related range-finding and dose-setting studies are extremely expensive (i.e., up to several million U.S. dollars/chemical), require large numbers of animals (i.e., about 2,000 animals per chemical) and are lengthy (i.e., 5–12 yr/chemical). Even though these studies are "gold standards" of the world, considering the approximately 600,000 chemicals in commerce the number of chemicals for which we have adequate toxicology information for risk assessment is minuscule.[66]

In a few notable cases, regulatory agencies have attempted to group chemicals for more efficient regulation, departing from the substance-by-substance approach for setting health standards. PCBs, which were regulated by Congress and banned in 1976 after 46 years in production, are defined as a class of over 200 closely related chlorinated synthetic organic compounds.[67] In 1973, the Occupational Safety and Health Administration (OSHA) issued emergency temporary standards, and ultimately permanent standards, regulating occupational exposures to a group of 14 carcinogens, representing chemicals used in the photographic and dye industries. The rulemaking and associated litigation took 40 months. Subsequently, OSHA proposed a set of regulatory actions on carcinogenicity based on four generic categories. For example, designation of a proven carcinogen was to be based on two positive animal tests or one positive animal test and positive evidence from short-term assays. OSHA would issue an emergency temporary standard for such chemicals. Such generic approaches to regulating chemicals spurred intense industry reaction and costly litigation, and eventually the agency's approach to regulating groups of chemicals was limited by the new deregulatory mood that prevailed over U.S. environmental policy in the 1980s.

More recently, bills supported by environmental groups have been introduced to phase out a large class of chlorinated organic compounds. Thus, the prospect of introducing a policy designed to regulate endocrine-disrupting compounds as a class (or classes) of chemicals is not unprecedented. Nevertheless, it is being fought vigorously by manufacturers who would prefer substance-by-substance rulemaking. One early study noted that chemical regulation moves at a snail's pace "because the cumbersome statutory procedures provide a chemical with a full panoply of due process rights accorded to any individual in our constitutional system."[68] Our system of jurisprudence, grounded on the principle of individual rights, has granted legitimacy to class action suits. Chemicals, as classes, may also gain standing in regulatory decisions, but so far that has been the exception rather than the rule.

Even with a dependable, inexpensive, and short-term assay that can identify the risks of xenobiotic hormone mimics or antagonists, a decision on how these chemicals should be regulated could easily become mired in the legal process. It probably will not be sufficient to show that these chemicals *can* disrupt reproduction; it must be demonstrated that current exposures and ambient levels pose an unreasonable risk to humans or wildlife. This may be accomplished either by demonstrating that the class or classes of chemicals are the cause of past reproductive or developmental anomalies based on epidemiological data or through controlled laboratory experiments on human cells or animals. However, for most of the chemicals currently in commercial use, the burden is on those who wish to restrict their usage to demonstrate a cause-and-effect relationship.

Another problem facing regulators is how they will address the by-products of chemicals that test negatively for endocrine-disrupting properties. Once emitted into the environment, these chemicals may recombine with other chemicals or degrade into products that are hormone mimics or antagonists. The government's burden of proof for regulating the commercial or industrial chemical sources of dangerous by-products has always been high. For example, nitrosamines, a potent class of carcinogens, are a by-product of the nitrites and nitrates that are used in meat processing. Sodium nitrate used as a preservative can react with amines in the stomach to produce nitrosamines. Regulators have argued that the benefits of the food additive in preventing botulism outweigh the risks of cancer. It is reasonable to assume that similar cost-benefit analyses will be applied to endocrine disrupters. Unless regulators have a means to assess the exposure level and potency of these chemicals in biological systems and their cumulative and combinatorial effects, the use of current laws to regulate them may face insuperable obstacles.

Regulators will also continue to be faced with the substitution dilemma, as they have been each time a product is banned or restricted. DDT and other pesticides developed in the 1940s and 1950s were replaced by chemicals that were less persistent in the environment but, in some cases, more acutely toxic. Barring the elimination or phasing out of pesticides altogether, regulators may see their role as being forced to choose between the lesser of two evils—a suspected carcinogen or an endocrine disrupter.

Restructuring the System of Chemical Regulation

In 1874 the widely circulated *Popular Science Monthly* heralded the new era of synthetic chemistry, citing as its goal the production of "artificially new compounds out of old materials."[69] By the last third of the nineteenth century, European chemical companies were synthesizing artificial dyes, perfumes, food flavorings, drugs, and pesticides. By the second half of the twentieth century, millions of new compounds had been created. Tens of thousands of synthetic organic chemicals have been introduced into commerce through agriculture and manufacturing, contributing to the unprecedented economic growth of the postwar period. Yet during this same interval, the industrialized nations have been floundering over how to regulate the synthetic molecules that continue to invade and threaten the habitats of humans and wildlife, even in the remotest places on earth.

It was during the 1960s and 1970s that the chemical revolution was subjected to its harshest criticism. At the time many believed that most cancers were caused by environmental carcinogens. The United States, followed by Europe and Japan, enacted a cascade of laws in hopes of screening and testing new chemicals entering the marketplace. Today, the confusion and divisiveness over regulating chemicals have never been more apparent. With the exception of tobacco, environmental chemicals are no longer seen as primary causes of cancer. Instead, sedentary lifestyle, obesity, smoking, genetics, and diet seem to be at the top of the list. With cancer increasingly understood as a multistage process, environmental carcinogens are viewed as one of many factors that might contribute to cancer risk.

After years of industry lobbying, U.S. policy has retreated from the 1958 Delaney clause, which prohibited any chemical that was found to cause cancer in humans or animals at any dose from being used in processed food. The newly passed Food Quality Protection Act of 1996 preempts the application of the Delaney concept to pesticide residues in food in favor of a "reasonable certainty of no harm" standard. Similarly, the cleanup of toxic waste sites is being viewed as overreactive and too expensive a process for society to continue to underwrite. Public reactions to residues of certain chemicals in food, such as the ripening agent Alar and the fungicide ethylene dibromide (EDB), have been attributed to irresponsible media hype. A new generation of journalists and environmental risk specialists is beginning to view synthetic chemicals as friends and not enemies and the plaintiffs against "chemical crimes" as the purveyors of junk science.

Yet just as the petrochemical industry was beginning to feel comfortable about reintroducing the word *chemical* into its corporate logos, a new theory of chemically induced diseases was beginning to weaken public confidence—again raising the worry budget of the world's largest chemical manufacturers.

As previously noted, the environmental endocrine hypothesis potentially implicates a broad range and significant production quantities of structurally diverse chemicals. In some respects the regulatory challenges posed by xenobiotics are similar to those for carcinogens, which also display wide variations in chemical structure. But in contrast to carcinogens, which have been at the epicenter of U.S. regulatory policy for decades, endocrine disrupters have just entered the scene. Although greater knowledge of the causal mechanisms of disease as mediated by endocrine disrupters would certainly aid regulators and the public health community, in the past chemicals have been restricted or banned even if such mechanisms were not fully understood. A case in point is asbestos, the use of which was dramatically restricted while scientists continued to debate how it caused lung cancer. For policymakers, the details of the causal mechanism are far less significant than some demonstration of causal association between the chemical and a disease.

The process of chemical risk assessment has been slow and ponderous. Toxicity information is estimated to be unavailable for some 80 percent of commercial chemicals.[70] The regulatory burden would be eased if the class of endocrine disrupters were small or if there were alternatives for the most serious offenders. Currently more than 70 chemicals (including chemical groups like PCBs, phthalates, and phenols) used substantially as pesticides and plasticizers have been implicated as being actual, probable, or potential endocrine disrupters. According to Wiles, more than 220 million pounds of pesticides known or suspected to be endocrine disrupters were applied to 68 different crops in recent years, with atrazine (classified by the EPA as a possible or probable human carcinogen) alone making up 29 percent of the total weight and 22 percent of the total acreage treated.[71] Even without making new demands of the pesticide regulatory process, the current system is widely recognized as inadequate to the task before it. Farmers, for example, are generally still accorded the right to use an effective pesticide. As Dorfman notes, "The restrictions that are appropriate for any pesticide depend on the availability and effectiveness of substitutes."[72]

It has become fashionable today to write about the need for restructuring environmental regulation. This trend is based on one or more of the conclusions that regulations are inefficient, irrational or illogical, unscientific, burdensome to industry, ineffective for protecting the public, and unresponsive to cost-benefit analysis. Many of the new critics of current environmental regulations support the conservative agenda that calls for downsizing federal regulations.

The emerging scientific evidence for endocrine-disrupting chemicals poses a challenge to the already overburdened regulatory systems of the world's industrialized nations. Most chemicals have not been adequately screened for reproductive, immunological, and developmental effects. If we choose to move beyond the current slow pace of progress in the regulation of hazardous chemicals, some changes will have to be made in the fundamental way that our society evaluates new and existing synthetic compounds. We will need to reach a consensus on effective and inexpensive assays for identifying endocrine disrupters and on methods for assessing the human health risk of cumulative doses. In the past, agencies used different action criteria for deciding whether to regulate or ban a chemical. Once assays have been developed and validated, they should be applied uniformly, based on some system of prioritization, to the over 75,500 chemicals used in industrial production and the nearly 900 pesticide active ingredients.

The standards of regulation for a chemical agent should not depend on how the chemical reaches the body but rather on the level of exposure and the biological properties of the chemical. Endocrine disrupters that enter the food chain as pesticides should be regulated like food additives (which are under the jurisdiction of the Food and Drug Administration) if their effects (e.g., reproductive toxicity) are comparable. Similarly, people may be exposed to endocrine disrupters in cosmetics or plastics, which are minimally regulated, at higher concentrations than other sources of the same chemicals. For a number of years there were two sets of de facto regulations governing foreign chemicals in food: one directed to additives in processed foods and another to pesticide residues in fresh produce. In 1990 the Natural Resources Defense Council joined the state of California in filing suit against the EPA, arguing that the agency had failed to apply the 1958 Delaney clause to dozens of pesticides that had been found to cause cancer in certain mammalian species in high-dose experiments. A settlement of the suit in 1994 was expected to remove 36 pesticides from commerce, but it left open both the future of the Delaney clause and its general application to pesticide residues.[73] In fact, the settlement became moot after the enactment of the Food Quality Protection Act in 1996, which finally resolved the contentious issue of carcinogenic pesticides to the satisfaction of some environmental and industrial stakeholders and established as a goal an approach to food safety based solely on health risks and not on the route of contamination or the products in which the contaminants are found.

In light of the pesticide debates, we need an effective way to address the endocrine disrupters that have become part of the waste stream through industrial effluent or pesticide runoff and that enter the food chain through packaging or the concentration of persistent organic chemicals in marine organisms. This will require a reexamination of the criteria for designating pollutants under the Clean Water Act, much of the emphasis of which has been on heavy metals, carcinogens, nitrates and phosphates, and more recently PCBs and dioxins. Efforts to reduce the total environmental load of endocrine disrupters may also require amendments to or new rulemaking under FIFRA if it is shown that pesticides are a significant source of human exposure to endocrine disrupters. Current information reveals a regulatory problem of immense scale, especially in the worst case, in which the chemicals most pervasively in use pose risks to humans or wildlife.

We must also begin to address the problem of transnational food shipments, through which endocrine disrupters may be introduced into U.S. markets at levels above those permitted in this country. Nearly a decade ago, a high-ranking EPA administrator acknowledged that a minuscule number of imported bananas treated with benomyl (classified as a probable endocrine disrupter [see Table 3] and a possible carcinogen) had been inspected at the Mexican border. U.S. border inspectors, who are supposed to monitor pesticide residues on foreign produce, simply cannot meet the growing demand for imports as funds for inspectional services decline and trading markets for agricultural goods expand. As Wargo points out, "Over 6 billion pounds of fruit and 8 billion pounds of vegetables are imported into the country annually. Yet FDA tests only eight thousand imported fruit and vegetable samples yearly, which means that on average roughly one residue test is performed for each 2 million pounds of imported food."[74]

The TSCA must be strengthened if PMN requirements are to include mandatory screening for endocrine disrupters. The reporting and inventory requirements under the TSCA should be amended to provide information on whether the chemicals currently in use are potentially endocrine disrupters in vertebrates. Some have argued that the TSCA rules must begin to look more like FIFRA, under which the burden is on the producers of chemicals to provide data that demonstrate that the uses to which the chemicals are put are well within safety limits for humans and the environment.

Finally, environmental regulations must address the issue of cumulative xenobiotic endocrine disrupters from multiple chemical sources. The accumulated or lifetime exposures of individuals to particular chemicals have been factored into risk assessment models.[75] But a system of regulation and risk assessment may have to take account of *total* xenobiotic chemical load originating from *different* agents if the additivity effect is confirmed by scientific experiments. This is a new challenge for regulators that will require an integrated look at chemicals. The closest analogy we have is radiation standards. Annual and lifetime exposure limits have been established for workers in the nuclear industry, and a common metric for all sources of ionizing radiation, along with the assumption that radiation risk is cumulative, makes this regulation possible. Some progress in this area is found in the Food Quality Protection Act, which is the first major piece of legislation that makes provision for cumulative exposures for all sources of chemical pesticides.

In summary, the current structure of environmental regulations, established in the 1970s, has been the target of criticism from every ideological direction. The courts, public pressure, and politics have been the primary determinants of environmental health priorities—leading to what Lave and Upton describe as disjointed, start-and-stop, frenetic regulation that takes a narrow view of problems and ignores altogether those that cross national boundaries.[76] Like the science that informs it, the process of social regulation has taken a reductionist approach, seeking chemical-by-chemical solutions; focusing on too few etiological outcomes; neglecting additive, cumulative, and synergistic effects; and allowing a balkanization of regulatory authority. A shift to a more systems-oriented approach to assessing health effects will require restructuring of the legal foundations and regulatory framework guiding the management of environmental chemicals.

Implementing a Screening Program for Endocrine Disrupters

Even while scientists were debating whether human diseases or developmental abnormalities had resulted directly or indirectly from endocrine disrupters, the U.S. Congress decided that it was not going to wait for a consensus on the human health risks before taking action. Congress enacted two laws with strong and directive language calling for immediate attention to the issue of identifying endocrine disrupters.

Under the Food Quality Protection Act and the reauthorization of the Safe Drinking Water Act, the administrator of the EPA, in consultation with the secretary of Health and Human Services, was required within two years of August 1998 to "develop a screening program, using appropriate validated test systems and other scientifically relevant information, to determine whether certain substances may have an effect in humans that is similar to an effect produced by naturally occurring estrogen, or such other endocrine effect as the Administrator may designate."[77]

The two laws focus on different sets of chemical substances and target different media of exposure. Section 304 of the Food Quality Protection Act authorizes the EPA administrator to test food-use pesticides or any substances that contribute to their effects if a substantial population is exposed, whereas Section 136 of the Safe Drinking Water Act amendments gives the administrator authority to test chemicals in drinking water. Other requirements under the legislation include a timetable of three years from the passage of the act for the implementation of a screening and testing program (due by August 1999), and one of four years for the EPA to deliver a progress report to Congress (due by August 2000).

Two offices of the EPA have responsibility for coordinating efforts to implement the new congressional mandate on endocrine disrupters. The Office of Prevention, Pesticides and Toxic Substances has primary responsibility for pesticides and industrial chemicals, while the Office of Safe Drinking Water is responsible for evaluating endocrine effects of drinking water contaminants. The EPA acknowledged the uncertainties about the health and environmental effects of endocrine disrupters in the background paper written in preparation for its implementation of the acts: "While we are working to answer these important scientific questions . . . EPA believes the potential implications of endocrine disruptors for our children and for our future are serious enough to warrant the agency taking prudent, preventive steps, without waiting for the research to be complete."[78]

The EPA announced it had already taken action to ban the use of a group of compounds with possible endocrine-disrupting effects, including PCBs, chlordane, DDT, aldrin and dieldrin, endrin, heptachlor, kepone, toxaphene, and 2,4,5-trichlorophenoxyacetic acid (2,4,5-T). Several of these regulatory initiatives were prompted by legal challenges filed against the agency by public interest groups. The agency also reported its plan to reassess four organochlorines currently on the U.S. market (dicofol, methoxychlor, lindane, and endosulfan) for their synergistic or cumulative effects. These chemicals are cited by the Illinois Environmental Protection Agency as known endocrine disrupters in animals or humans (see Table 3).

In anticipation that Congress would incorporate testing for endocrine disrupters into its amendments of FIFRA and the Safe Drinking Water Act, the EPA held a stakeholder meeting on May 15–16, 1996, to determine how the agency could work cooperatively with industry, the environmental community, and academia to develop a screening and testing strategy. On October 16, 1996, about two months after the passage of the legislation calling for a screening and testing program, the EPA chartered the Endocrine Disrupter Screening and Testing Advisory Committee, more commonly referred to by its acronym EDSTAC. The committee was given a two-year life and a budget of $1.25 million to provide guidance to the EPA on the establishment of a comprehensive screening and testing program for pesticides and other chemicals.

Reflecting the pressure on the EPA to meet the congressional timeline, the EDSTAC met 9 times and convened approximately 15 additional working committee meetings between plenary sessions over a 20-month period. The EPA convened a public organizational meeting on October 31–November 1, 1996, that included approximately 30 people who had been nominated for appointment to the committee and 200 other members of the public. This meeting (as well as subsequent ones) was facilitated by the Keystone Center, a nonprofit consensus-building organization, to address how the EDSTAC would be organized. One of the first issues discussed was whether the committee should be a freestanding federal advisory body under the auspices of the Federal Advisory Committee Act or whether it should be piggybacked onto a pre-existing committee created under the act. The Clinton administration, like its predecessor, had been using the advisory committee act mechanism liberally (in some cases because of statutory requirements), and as a result it now faced budgetary limits on the number of such committees that would be approved by the Office of Management and Budget. Committees created under the Federal Advisory Committee Act operate under special statutory requirements, including open hearings, broad stakeholder participation, and liberal public notification provisions that grew out of post-Watergate government "sunshine" laws. As a result, they can cost more to implement than more garden-variety types.

EDSTAC members-to-be voiced strong preference for a full committee charter. Otherwise, they argued, the committee's recommendations would have to be filtered through another administrative layer and thus be subject to a parent committee's veto power. The EPA saw this as an important enough issue to argue strongly for a new committee charter under the Federal Advisory Committee Act, which was eventually approved by the Office of Management and Budget.

According to its charter, the EDSTAC was to be composed of diverse groups of individuals representing a broad range of interests and backgrounds. As of January 1997, the committee had 48 members and alternates, including 4 from EPA; 7 from other federal agencies; 3 from the state environmental protection agencies of New York, California, and Wisconsin; 8 from industry; 2 from worker protection and labor organizations; 8 from public interest groups; and 6 from universities. Membership on the committee fluctuated between 39 and 48 members and alternates. Some members were added, some withdrew in midstream, and some remained inactive.

The committee was given several objectives:

— To develop a flexible process to select and prioritize chemicals and pesticides for screening.
— To develop a process for identifying new and existing screening tests and establish mechanisms for their validation and procedures for their early application.
— To establish a process and criteria for deciding when tests beyond screening will be used and how they will be validated.
— To communicate its decisions and recommendations clearly to the public.

One of the EDSTAC's earliest objectives was to establish the scope of the screening and testing program. It decided that the program should focus on three types of hormones (estrogens, androgens, and thyroid-related hormones), that it should study two types of effects (agonist and antagonist), and that it should not limit the effects it would consider to those of chemicals that bind to hormone receptors (one mechanism of endocrine disruption).

Neither the laws nor the EDSTAC itself had anything to say about what would happen to those chemicals that were found to disrupt, to some degree, the endocrine systems of humans or wildlife, although the Food Quality Protection Act and the 1996 Safe Drinking Water Act amendments provide new testing authority. Without this new legislation, the EPA would have had to fall back on its traditional authority under the TSCA, FIFRA, the Clean Water Act, and the Safe Drinking Water Act to regulate endocrine disrupters. The EPA's legislative authority varies significantly under these acts. For example, the EPA does have authority to order testing of chemicals under the TSCA, yet the weakness of its enforcement authority is evident. Since its passage in 1976, this authority has been used for only 121 chemicals—a mere 0.2 percent of the TSCA chemical inventory.

With the strict deadlines mandated by the legislation, the EPA had laid down a considerable challenge to the committee members, many of whom were not accustomed to working toward a common goal across such divergent environmental interests. The EDSTAC meetings were chaired by Lynn Goldman, a pediatrician and public health specialist who, as director of the Office of Prevention, Pesticides and Toxic Substances, was among EPA administrator Carol Browner's most trusted appointees.

Goldman acted strategically to establish her own priorities before the committee. She asked the members to develop consensus-based recommendations on a screening and testing strategy that was scientifically defensible. However, the diversity of interests represented by the group posed a significant obstacle to reaching a consensus. In April 1997, between the third and fourth plenary meetings, Goldman responded to questions about the possibility of a minority report, clarifying her goals and underscoring the importance of unanimity on some key issues. In a letter to the EDSTAC members, she wrote:

> I see the major advantages of a consensus being that all the major stakeholder groups will know that their views and interests were balanced against the views and interests of others—everyone is equal on the committee. I also hope that a consensus outcome will translate into a voluntary effort on the part of affected industries to implement the recommendations of the EDSTAC even in advance of their final implementation by EPA. We also believe that a consen-

> sus of all major stakeholders will minimize, and hopefully eliminate, the likelihood of any court challenge to the screening and testing program that is developed on the basis of the EDSTAC's recommendations.[79]

The consensus strategy requested by Goldman meant that the EDSTAC participants would have to subordinate their individual ideological positions in order to seek a middle ground. Environmental activists sat with industry representatives with the understanding that their task was to find a practical way to meet the congressional mandate of a screening and testing program for endocrine disrupters. Grandstanding for the cause of environmental purity or economic prosperity was negatively reinforced in favor of achieving compromise solutions.

The EPA had been given a task and a timetable that many seasoned observers of environmental regulation could only describe as daunting and impractical. Less weighty challenges to the agency, such as pesticide reregistration, had fallen years behind schedule. In the case of testing chemicals for their endocrine-disrupting properties, not only was the EPA's congressionally mandated timetable short, but the task is unusually complex. First, there is the question of how many distinct chemicals will fall under the testing program. The figure for all chemicals in consumer products, agriculture, and industry could easily reach 100,000. As has been noted, some 75,500 chemicals are listed under the TSCA inventory. In addition, as of 1998, the EPA had registered 884 pesticide active ingredients and permitted 2,500 inert pesticide ingredients to be used in the formulation of over 24,000 pesticides. Then there are the approximately 8,000 chemicals, regulated by the Food and Drug Administration, used in cosmetics and food additives, as well as unregulated nutritional supplements and natural substances such as phytoestrogens and mycotoxins, some of which may also exhibit endocrine effects. To illustrate the magnitude of the task, if a battery of tests for a single chemical takes five days to execute, and if a new chemical test is completed every five days, it could take over a thousand years to complete the testing program.

In addition to the number of distinct chemical agents, there is also the problem of chemical mixtures. Testing each chemical by itself may underestimate the additive or possible synergistic effects of chronic exposure to scores of compounds. If the EPA is to address mixtures among the estimated 87,000 chemicals to be analyzed, the effort would involve a level of complexity that could easily overwhelm our most advanced testing systems and surely our federal budget. Howard calculates that to test only the commonest 1,000 toxic chemicals in unique combinations of three at a single dose per experiment would require 166 million different experiments.[80]

The sheer number of chemicals suggests that the EPA must establish priorities for chemicals that should be tested. Here it is faced with a number of competing criteria. Should the agency select the chemicals produced in greatest quantity? Or should it be concerned about the chemicals to which humans (or wildlife, or both) are most often exposed in their daily lives? Should specific types of exposures (e.g., those occurring via food, water, or air) be emphasized first? Should there be exemptions of chemicals that have a low probability of having endocrine effects? Should the agency give priority to the chemicals for which there is already evidence of endocrine-disrupting potential? In connection with this last potential criterion, endocrine-effects data are lacking for most chemical substances. In 1998 it was estimated by the EPA that at least some developmental and reproductive toxicity screening and testing data were available in the published literature for only about 5,000 chemicals, most of which are pesticides and pharmaceuticals. How can the EPA establish testing priorities based on available data with such a paucity of information? It may have taken decades to accumulate the reproductive and developmental toxicity data for those 5,000 chemicals. One of the challenges to the EDSTAC was to speed up the process of generating the relevant data to set priorities for risk assessment of endocrine disrupters.

The EPA was also faced with the decision of whether to focus its screening and testing exclusively on human health impacts of endocrine disrupters or also to consider the health of wildlife. There was a logic to directing the program to human health effects, at least initially. The laws that had established the testing program (the Food Quality Protection Act and the amendments to the Safe Drinking Water Act) are directed at human health and not at more general environmental concerns. In addition, by holding off on environmental effects and giving priority to effects in humans, the screening and testing program could be vastly simplified, since it was uncertain how many different animal species would have to be tested to achieve a full range of environmental impacts. It had been the EPA's decision to include fish and wildlife and invertebrate biologists on the EDSTAC, reflecting the broader potential scope of endocrine effects.

Among the more contentious issues raised within the committee was *risk assessment.* Some committee members made it a condition for their participation that the decisions arrived at regarding screening and testing strategy not imply that they favored a risk assessment methodology as a risk management tool. The use of risk assessment in setting chemical standards had become a deeply divisive and ideological issue. Those who viewed themselves as progressive environmentalists were critical of such uses of risk assessment because they considered the field of toxicological risk assessment unduly influenced by industry and believed that technical risk assessment was often used to discredit democratic processes.

Under the procedure followed by the EPA, the EDSTAC delivers its recommendations to the administrator of the agency; the recommendations are reviewed by a scientific advisory panel set up by the agency for this process; modifications are made; the agency formulates its plan; the draft plan is published in the *Federal Register* for public input; a joint panel consisting of FIFRA's Scientific Advisory Panel and EPA's Science Advisory Board reviews the plan and the public comments and offers its recommendations; and, based on scientific and public comments, the EPA administrator issues the final screening and testing plan. Historically, the Science Advisory Panel has almost exclusively addressed issues associated with FIFRA while the Science Advisory Board has addressed virtually all other science issues that arise at the EPA. With respect to endocrine disrupters, which touch on several regulatory roles of the agency, a joint SAP/SAB panel was established, with designated federal officials from each, to review the EDSTAC's recommendations.

However, before the EDSTAC could even begin to tackle the complex technical issues, many involving state-of-the-art science, the committee recognized the importance of developing an overall conceptual framework that would establish the general goals and boundary conditions of the program and set the architecture within which the details of screening and testing could be worked out.

One of the thornier conceptual issues that divided the members was the very definition of an endocrine disrupter. Initially, the committee used a working definition derived from an EPA-sponsored workshop,[81] which described an endocrine disrupter as "an exogenous agent which interferes with the synthesis, secretion, transport, binding, actions, or elimination of natural hormones in the body which are responsible for the maintenance of homeostasis, reproduction, development or behavior."[82] However, some members of the committee were dissatisfied with the definition, calling it too open-ended for use in a regulatory context. They argued that there might be some detectable interference or modulation of natural hormones, but that the effects on the life of the organism might be minimal. By May 1997, the EDSTAC had adopted a second working definition, which included terms like *adverse effects* and *progeny* and introduced population effects: "An exogenous substance that changes endocrine function and causes adverse effects at the level of the organism, its progeny, and/or (sub)populations of organisms."[83]

Although the new definition was useful in getting the committee to make progress in the design of the screening and testing program, some members remained opposed to the term *adverse effects*. They argued that chemicals might induce abnormal fluctuations in hormone levels without any observable adverse effects, or that the adverse effects of exposure to xenobiotics might take years to show up in the organism. The use of the term *adverse effects* would, according to this view, impose an unduly high standard for calling a chemical an "endocrine disrupter." Those who were adamant about including the language *adverse effects* countered that the definition should distinguish "endocrine disruption" from a wide range of chemically induced hormone fluctuations to which the organism could adapt. To achieve consensus, the EDSTAC agreed not to define an "endocrine disrupter" but instead to adopt a general description that would express the range of views of the committee's members: "The EDSTAC describes an endocrine disrupter as an exogenous chemical substance or mixture that alters the structure or function(s) of the endocrine systems and causes adverse effects at the level of the organism, its progeny, populations, or subpopulations of organisms, based on scientific principles, data, the weight-of-evidence approach [discussed later in this chapter], and the precautionary principle."[84]

The strategy pursued by the EDSTAC was to create an initially wide lens for determining which chemicals qualified for screening and testing and to narrow the lens as more information about a given group of chemicals was acquired. The underlying assumption of this approach was that, barring evidence to the contrary, chemicals were guilty of being potential endocrine disrupters unless proven innocent. The committee agreed that the screening and testing program for endocrine disrupters should be directed at both human health and ecological effects, notwithstanding the human health orientation of the laws mandating the program. It also proposed that the scope of the program not be limited to the estrogenic effects of chemicals, but that it also incorporate other endocrine-disrupting effects.

The EDSTAC considered both a short-term and a long-term goal for the screening and testing program. For the short term, the committee proposed that the program identify and characterize the risks of endocrine disrupters that enhance, mimic, or inhibit estrogenic, androgenic, and thyroid-related processes, three commonly studied categories of endocrine effects. Yet it recognized that this plan did not cover all the hormone-mediated pathways that could be disturbed by foreign chemical messengers. Therefore, as a longer-term goal the EDSTAC recommended that other hormonal effects be considered, including chemical interference with hormones produced in the brain, such as prolactins (those involved with the production of breast milk) and luteinizing hormone (which initiates ovulation in females and testosterone production in males).

The committee also recommended that the EPA consider tests that would detect multiple hormone interactions (recognizing the complex signaling patterns of the endocrine system, through which chemicals can react through receptor-binding and neurobehavioral end points), examine multiple species, and consider long-term and delayed effects. In addition, the committee's conceptual framework included the proposal that the screening and testing program examine both chemical substances and common mixtures and that it should be capable of ascertaining the additive, synergistic, and antagonistic effects caused by the interactions among components in a mixture.

The EDSTAC's ever-widening lens for screening chemicals also included naturally occurring nonsteroidal estrogens, such as phytoestrogens (plant estrogens) and estrogenic mycotoxins (chemicals produced by fungi). Why would the committee suggest screening for their estrogenic properties plant products that have been an established part of the human diet? Some members of the committee felt that the analysis of these substances would enable the EPA to validate the testing protocols for synthetic compounds. The assays, they argued, should be able to measure synthetic as well as naturally occurring estrogenic substances. In addition, phytoestrogens can be used to quantify and compare the relative potency of estrogenic chemicals, both naturally occurring and synthetic.

Some industry representatives on the EDSTAC contended that it was unfair to regulate chemicals by their origin (e.g., synthetic) rather than by their effects. One of the arguments disputing the claim that synthetic estrogenic substances can induce human health effects is based on the relative potency of foreign and natural estrogens: if the food we eat contains much greater quantities of estrogenic compounds than the synthetic chemicals that enter our bodies, what grounds are there to be concerned about the latter's estrogenic effects? Because the synthetic estrogenic compounds may have other distinguishing properties (e.g., they do not metabolize quickly or they affect other hormone receptors), some means of comparing nonsteroidal estrogens and synthetic xenoestrogens would seem important in discussing the human health risks of endocrine disrupters.

Chemical mixtures always present a problem for regulatory agencies. Which mixtures does one choose to study and in what proportion of compounds? The EDSTAC was all too aware that by introducing mixtures into its screening program it had created the potential for overloading the system. Taking a pragmatic approach, the committee placed priority on six classes of chemical mixtures, considering the high degree of human exposure to the compounds, with particular attention given to infants and children:

1. Contaminants in human breast milk.
2. Phytoestrogens in soy-based infant formula.
3. Mixtures of chemicals most commonly found at hazardous waste sites.
4. Pesticide/fertilizer mixtures commonly detected in surface water and groundwater across the United States.
5. Disinfection by-products, especially chemicals used in purifying drinking water.
6. Gasoline, which contains a complex mixture of volatile organic compounds.

The by-products of chlorination in drinking water have been an ongoing concern to the public health community, which does still consider chlorination to have an overall positive risk-benefit ratio. Nevertheless, new studies employing increasingly powerful statistical methods are finding adverse health effects of chlorine by-products. For example, in a prospective study of three regions of California, Swan et al. reported high spontaneous abortion rates among women who drank more tap water than bottled water.[85] The suspected chemical culprits are trihalomethanes—one of the principal groups of chlorination by-products detected in California drinking water. Screening and testing the chlorine by-products for endocrine-disrupting effects could provide clues to the mechanisms underlying the epidemiological findings of increased spontaneous abortions.

At its October 1997 meeting, the EDSTAC debated whether it should construct a list of high-priority chemicals for screening. It ultimately rejected that idea and instead endorsed the concept of a nominating process that would enable citizens to recommend specific chemicals for screening of endocrine effects. Some industry members of the committee felt uneasy about developing a list of high-priority chemicals. They argued that such a list might be interpreted as a conflict of interest for those members who worked for chemical companies, since some of the chemicals on it might be manufactured by their competitors. There were also concerns among industry spokespersons that the mere appearance of any chemical on a list—even if it were labeled a "*potential* endocrine disrupter"—could stigmatize the product before all the information and testing results were available.

The EPA's Initial Plan

Thus the EDSTAC—some 40 individuals representing many different sectors meeting over a two-year period—reached a consensus on a plan to determine which chemicals should be evaluated for their endocrine-disrupting properties, by which processes, in what order, and by which screening and testing protocols. The scope of the screening and testing program they ultimately proposed was unprecedented in U.S. regulatory history. Unfortunately, the program does not lend itself to a simplified narrative description, and even its graphic visualization (Figure 2) belies its complexity. In electing to emphasize the key elements of the plan, I will inevitably gloss over some of that complexity.

First, it should be emphasized that this is not a linear plan under which a chemical is taken through a series of tests—like a car moving through a car wash—until a final assessment of its risk level is made. Rather, some chemicals are taken through the system faster than others. Other chemicals are placed in holding patterns until more information is available.

Second, under the EDSTAC proposal no chemical is ever free from suspicion as an endocrine disrupter. Some chemicals will be taken through the full battery of tests and sent for hazard assessment. Others that do not reach the final stage of assessment will be kept under surveillance. However, there is no provision in the plan for labeling a chemical "free of endocrine-disrupting properties" unless actual data are developed to support this contention. The EDSTAC established a "hold" status for chemicals to account for the possibility that emerging scientific techniques and new information could shed new light on the priority for testing compounds previously considered at low risk for endocrine disruption.

The EDSTAC screening and testing program may be divided into five stages. Each stage provides different options for screening and testing based on what is already known or what is learned about chemicals as they pass through the process. The stages are (1) sorting of chemicals, (2) priority setting, (3) screening, (4) testing, and (5) hazard assessment.

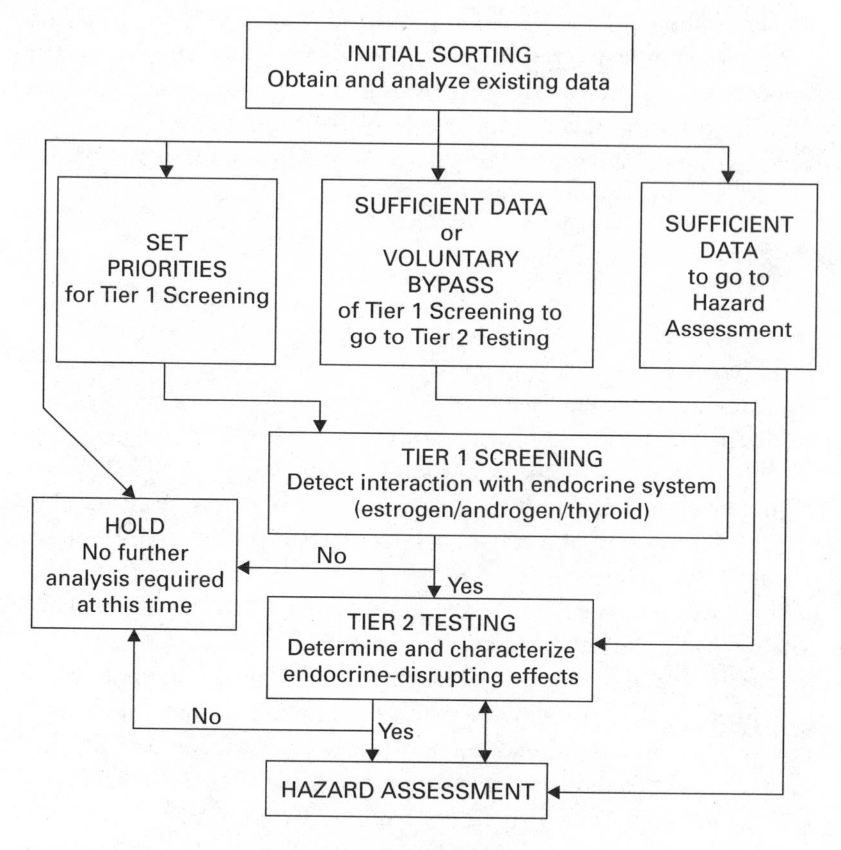

Figure 2
EDSTAC conceptual framework for screening and testing endocrine disrupters
***Source:* Endocrine Disrupter Screening and Testing Advisory Committee (1998)**

The first stage sorts the vast quantities of chemicals in commerce based on existing information and data into four categories, indicating whether the chemicals should be placed on hold or be introduced into one of the other stages for screening and testing. This is a necessary first step since limitations on research resources and laboratory capacity will not permit the simultaneous entry of hundreds to thousands of chemicals into the screening and testing program. The second stage, priority setting, establishes a phased approach toward screening and testing by assigning chemicals to the categories highest, medium, and lowest priority.

In the third or screening stage (referred to in the EDSTAC lexicon as Tier 1 Screening or T1S), chemicals face a battery of in vitro and in vivo (animal or embryo or both) assays designed to detect whether they will interact with the endocrine system, specifically to modify estrogenic, androgenic, or thyroid function.

A chemical that tests positive in the screening stage would qualify for the testing stage (referred to as Tier 2 Testing or T2T). At the Baltimore EDSTAC meeting held in March 1998, industry representatives were steadfastly against using the term *potential endocrine disrupter* for a chemical that has triggered some but not all of the screening and testing protocols. As a result, the term was eliminated from the draft report. Indeed industry representatives were opposed to any use of the term *disrupter* for a chemical unless it met the full panoply of requirements for the screening and testing process. For chemicals that tested positive in the screening phase (T1S), they proposed using the term *endocrine-active substance.* They argued that a chemical tagged an "endocrine disrupter," whether potential or actual, would be stigmatized—in effect found guilty before all the tests had been completed.

In the testing stage (T2T), long-term, whole-animal studies are used to determine the concentration of a chemical that elicits an endocrine response as well as any pathological effects. After completing testing, some chemicals will undergo a hazard assessment, which integrates all the available toxicity data and information; others will enter a holding pattern.

Let us now take a closer look at the five stages.

Stage 1: Sorting of Chemicals. Beginning with a universe of chemicals approximating 87,000, the EDSTAC proposed an initial sorting into four general categories: chemicals classified as polymers (molecules composed of repetitions of a single structure or monomer); chemicals for which insufficient data are available to assess their endocrine effects; chemicals for which there is sufficient data on endocrine effects to bypass screening and go directly to testing; and finally chemicals whose endocrine-disrupting effects are well enough understood to bypass all screening and testing and go directly to hazard assessment.

Why were polymers given a special category in the screening and testing plan? Simply because the EDSTAC was seeking ways to reduce the number of chemicals submitted for priority screening and testing so that the system would not become overloaded. Early in the deliberations, some committee members had suggested setting aside polymers on the assumption that their large molecular mass would keep them from being absorbed in the gastrointestinal tract. By excluding polymers from the plan, its backlog would be lightened by approximately 25,000 chemicals. But after it was reported by members of the committee that, in their first few months, newborns do absorb large macromolecules nonselectively through their intestinal tracts, the EDSTAC was persuaded to reconsider the exemption. Eventually, the committee chose not to recommend a blanket exclusion of polymers from the screening and testing protocols.[86] Instead it drew a distinction between new and old polymers and between the large macromolecules and their components.

Under its TSCA rules the EPA had already exempted polymers of molecular weight greater than 10,000 daltons (the dalton is a unit of molecular weight) from premanufacture notification with the rationale that chemicals of this size neither are absorbed into nor interact with biological systems. Under a consensus reached by the EDSTAC, 1,000 daltons was chosen as the threshold molecular weight for exempting polymers. Polymers greater than or equal to that weight would be evaluated by their components (e.g., monomers and oligomers, which are compound molecules made up of two or more monomers) and placed in a holding pattern. The smaller polymers (those with weights less that 1,000 daltons) would proceed to priority setting for screening.

Stage 2: Priority Setting. During their deliberations, members of the EDSTAC recognized that developmental and reproductive toxicity data were available in the literature for a relatively small percentage of the total number of chemicals, mostly pesticides and pharmaceuticals. Since "endocrine-mediated effects" had never been a general category in previous regulatory assessments of chemicals, very little would be gained in terms of accelerating the screening and testing process by accessing existing toxicological databases. The prospect of generating relevant risk information from traditional toxicological studies in a timely fashion (i.e., within 10 years) was daunting. The use of some method of priority setting and phased testing would be necessary if the system were not to be overloaded.

To aid in its priority setting, the EDSTAC turned its attention to automated screening methods developed over the past several decades by pharmaceutical and chemical companies to identify promising or problematic drugs and industrial chemicals before they entered clinical or experimental trials. The term *high-throughput prescreening* (HTPS) was used by the EDSTAC to refer to automated processes for chemical prescreening. Since the prescreening tasks could be accomplished mechanically by robots, the tests could run continuously, and large numbers of chemicals could be processed within a relatively short time. Although the automated tests would be limited in the range of data they could generate, their primary function would be to identify any chemical that has an affinity for the estrogen, androgen, or thyroid hormone receptors, that is, to determine whether the chemical binds or does not bind to the hormone receptor of interest.

The HTPS results were not considered as offering definitive evidence that a chemical does or does not have endocrine-disrupting properties. For one thing, the HTPS methodology covers only one type of mechanism by which chemicals can affect the endocrine system, namely, biological activity resulting from the binding of a chemical to a hormone receptor. There are other known mechanisms by which chemicals can interfere with hormonal signals. For this reason, HTPS serves a triage function, enabling regulators to determine which chemicals should be sent for further screening and testing. The EDSTAC recommended that approximately 15,000 chemicals, produced in amounts greater than 10,000 pounds per year, be screened by HTPS. Once screened, those that tested positive would be considered for the next screening stage, while those that tested negative would be placed in a "wait and see" category.

Stage 3: Screening. In the stage referred to as screening (in EDSTAC terminology Tier 1 Screening or T1S) chemicals would face a battery of in vitro and in vivo assays designed to detect their potential to interact and disrupt an endocrine system. The purpose of the screening program is to determine whether a chemical may have an effect similar to that of naturally occurring hormones and to identify, characterize, and quantify those effects for estrogen, androgen, and thyroid hormones. The in vitro screens involve special cell constructs, such as yeast cells with human hormone receptors inserted, which can test chemicals for a particular mode of action. Will the chemical bind to an estrogen, androgen, or thyroid receptor in the presence of certain enzymes? These tests are sufficiently specific and sensitive to yield many false positive readings. Simply stated, substances that bind to a hormone receptor in a cell-free preparation (that is, a preparation consisting of chemicals isolated from a cell) or even a whole cell may not cause an adverse biological effect in an intact animal.

The benefit of using whole-animal systems in addition to cell cultures is that whole animals provide a comprehensive evaluation of the endocrine system of the organism that is exposed to the foreign chemical, and they can therefore be used to study a broader range of mechanisms than in vitro assays. Of course, short-term animal tests may not pick up biological effects that would be manifest in longer-term animal studies. Longer-term studies also enable investigators to explore the effects of a chemical at critical life stages in the organism's development and to apply different doses and alternative routes of exposure.

In addition to the cell culture receptor-binding assays in the screening program, the EDSTAC proposed a 3- to 20-day rodent assay, a frog metamorphosis assay, and a fish gonadal assay to address whole-animal exposures at critical developmental windows. A chemical that tested negative in the T1S stage would be set aside in a holding category unless new statutes required periodic review or rescreening or a second look was otherwise mandated by new information. Alternatively, a chemical that tested positive for screening would be considered to have high priority for the next stage, called testing.

Stage 4: Testing. The testing stage (also called Tier 2 Testing or T2T) of the program would provide more comprehensive tests for endocrine disruption in animals. The battery of whole-animal tests would be used to determine the concentration of a chemical that elicits an endocrine response for estrogen-, androgen-, and thyroid-mediated processes in the whole organism. In addition to rodents, the species of choice in carcinogen testing, the EDSTAC recommended that four other, nonmammalian, taxonomic groups be used in whole-animal tests in T2T: birds, amphibians, fish, and invertebrates. A chemical that tested negatively in this battery of tests would be placed in a hold category and remain there unless there were reason for retesting. A chemical that tested positive would be sent to the final stage in the program, namely, hazard assessment.

The EDSTAC envisioned the battery of tests as involving longer-term studies encompassing critical life stages and processes, a broad range of doses, and the administration of test chemicals via relevant routes of exposure (e.g., air, water, food). The tests would be designed to provide a comprehensive profile of effects, including second-generation effects from in utero exposure. Although less sensitive than the screening battery, the tests would be more reflective of the actual conditions under which chemicals behave in the organism (absorption, metabolism, excretion) and more responsive to the full range of the organism's life cycle (growth, development, behavior).

Stage 5: Hazard Assessment. The final stage in the screening and testing program developed by the EDSTAC is the hazard assessment of a subset of chemicals that have completed the testing protocols in whole animals. Hazard assessment or hazard identification is one of the key steps in the risk assessment of a chemical agent. According to the conceptual framework for risk assessment published by the National Research Council, the four stages of risk assessment are hazard identification (the determination of whether a chemical is causally linked to a health effect); dose-response assessment (the study of the relationship between the magnitude of the exposure and the occurrence of the health effect); exposure assessment (the determination of the extent of human or wildlife exposure); and risk characterization (the integration of the prior three steps into an estimate of the magnitude of the risk posed to humans or wildlife).[87]

The proposed system for screening and testing endocrine-disrupting chemicals is ponderous, complex, and replete with ambiguities—some of which will presumably be resolved when the details of the program have been laid out. The issues surrounding the testing of potential endocrine disrupters seem far more complex than those regarding carcinogen testing. In the latter process, investigators were looking for tumor production in animal studies or mutations in cell studies. Some regulations called for replicated animal studies in two species. The end points were reasonably clear, namely, the production of tumors. The Delaney clause did not specify dose-response requirements. Regulatory agencies used highest tolerable doses or LD_{50} (the lowest dose that kills on average 50 percent of the experimental population) to help set the range of doses used to test for carcinogenicity of chemicals. Low-dose effects were extrapolated from high-dose experimental data.

In contrast, the proposed testing of endocrine disrupters includes multiple end points. Foreign chemicals may interfere with numerous hormone message pathways, resulting in an undetermined number of possible adverse outcomes. In estrogen receptor-binding assays, an investigator examines the ability of a chemical to compete with standard estrogenic chemicals, such as DES or estradiol. In other cases the investigator may be seeking to determine whether a chemical causes cell proliferation in comparison with a standard estrogen or interferes with thyroid hormones. There are also situations in which foreign chemicals initiate protein synthesis (so-called transcription activation events) by activating specific genes in the cell nucleus. Alternatively, scientists may wish to investigate a chemical's capacity to inhibit hormone synthesis. The xenobiotic may interfere with certain enzymatic pathways that are essential to the synthesis of hormones. The blockage of hormone synthesis, in turn, can result in a variety of adverse effects in the organism, including abnormal serum hormone levels, pregnancy loss, and interference with normal male and female mating behavior. Among the multiple outcome measures for in vivo tests, which have been in use for some 70 years, are uterine size, vaginal cornification, and age of onset for puberty. In avian studies there are several end points for investigating endocrine activity, including number of eggs laid, proportion of cracked eggs, eggshell thickness, and number of viable embryos and chicks surviving to 14 days.

One complicating factor in identifying endocrine disrupters is that some effects may manifest themselves only during certain stages of development. For other effects, a multigenerational test may be warranted. This is considered the most comprehensive test available for assessing reproductive and developmental toxicity. However, such tests cost between $350,000 to $800,000 per chemical—which explains their limited use.

Another problem facing regulators is how to evaluate the evidence from the screening and testing protocols. Suppose there are some positive and negative results from a given group of tests. How will these mixed results inform policy? The EDSTAC proposed a weight-of-evidence approach as the criterion for deciding whether a substance should proceed from screening to testing and which chemicals should be designated as endocrine disrupters after testing and moved to hazard assessment—the final stage before regulatory controls can be imposed. Weight of evidence is useful in reaching a judgment when one is confronted with inconsistent results for a substance across multiple assays involving different organisms. In lieu of prescribing a particular weight-of-evidence approach, of which there can be many, the EDSTAC agreed to offer some guidelines and allow the EPA to work out the details. These guidelines require an examination of all the evidence but place different values and significance on certain outcomes. In determining the weight of evidence, consideration is given to the number and types of assays that yield positive results, greater concern is attached to relatively low-dose effects, and endocrine disrupter effects from in vivo assays are given greater weight than in vitro assays.[88]

The weight-of-evidence approach to regulation, as adopted by the EDSTAC, means that there is no simple algorithm for ascertaining whether a chemical should be sent for hazard assessment. It is therefore very different from the criterion by which positive tumor indications on two different species automatically trigger a regulatory response to a suspected carcinogen. Although its decision process is highly transparent, the EDSTAC's proposed screening and testing program will not be easy for citizens—or even dedicated environmental stakeholders—to follow or understand. It is unlikely, based on its early description, that anyone—with the exception of certain individuals from industry or the public interest sector possessing quite specialized knowledge—would be able to keep abreast of, and offer intelligent comments and critiques on, the testing program.

It is noteworthy that this major governmental initiative, to create a toxicological inventory of industrial chemicals in response to a relatively new public health concern, was largely the result of breast cancer activists building their case around the scientific work of a woman of unusual foresight and initiative. Theo Colborn advanced the environmental endocrine hypothesis at a time when breast cancer activists were beginning to organize a mass movement to strengthen national priorities in searching for the cause(s) of the disease.

The screening and testing program that EDSTAC has proposed is the type of effort whose ongoing management is best suited to a centralized governmental agency such as the EPA. Over the years, the EPA has been mandated to take on other large-scale projects, such as the federal Superfund program, the maintenance of an inventory of all chemicals used in industry, the implementation of national air and water standards, and the re-registration of pesticides. The agency plans to use its authority under existing legislation to require industry to supply data on endocrine disrupters based on the tests that the government has validated for the task. Where it does not have a legislative mandate, the EPA can still seek enforceable consent agreements for screening high-priority chemicals.

Other government agencies have begun to support this effort. In 1996 the National Institute of Environmental Health Sciences and the Centers for Disease Control began collaborating on a study of human exposure assessment for environmental endocrine disrupters. Funded by the institute at about $700,000 a year, the centers began examining blood and urine samples from approximately 200 people for the presence of some 30 different compounds, including synthetic endocrine-disrupting compounds and phytoestrogens.

The problem of developing high-quality toxicological data on a broad array of effects across species may seem intractable to some, especially given the mixed history of large-scale regulatory programs. Others will look at the Human Genome Project and ask why similar progress on a large scale cannot be expected in the areas of ecological health and prevention. Still others, looking at that same project, will ask whether a comprehensive approach to assessing the risks of environmental chemicals is worth the cost. Whereas the government's role in the human genome initiative is to create opportunities for enhanced well-being and expanded commerce, its role in testing for endocrine disrupters is to determine whether some of the costs of commerce, at least in the chemical sector, have been too high. Perhaps that explains the difference in public perception. We have become accustomed to devoting significant social resources to create new wealth—but we are more hesitant to spend resources on a similar scale if they may ultimately yield discoveries that might constrain economic development.

CHAPTER 5

Conclusion: Expanding Paradigms of Chemical Hazards

Science plays a central role in shaping our images of the world, including our ideas about risk. For at least half of the twentieth century, the public's perception of chemical hazards has been influenced largely by the fear of cancer. The discovery that chemicals cause genetic mutations in somatic cells, along with the widely shared view that mutations are linked to the development of neoplasms (cancer cells), provided the scientific grounding for much of the social response to the class of chemicals labeled "carcinogens." Yet it was not until the early 1990s—when Theo Colborn, in collaboration with other scientists, publicized the broad outlines of a hypothesis that postulated a link between synthetic chemicals and a variety of reproductive, behavioral, and developmental abnormalities—that there was a serious reconsideration of the social priorities that had been established for evaluating the risks of human and wildlife exposures to synthetic organic chemicals released into the environment.

The cancer paradigm for assessing chemical risks had not resulted in the banning or restriction of many agricultural and industrial compounds.

Even when strong public intuitions implicated chemicals as the cause of elevated cancer rates, scientific studies were rarely able to confirm these intuitions.

The environmental endocrine hypothesis emerged at a time when the public debate over carcinogens had been waning and when the media was becoming blasé about chemical risk stories. In early January 1988, the *New York Times* published an editorial, titled "Worry Chic," expressing a ho-hum view about the world of environmental concerns: "Don't relax. There's always something to worry about. . . . There may be urethane in the wine. There may be parasites in the sushi. . . . Even if wine has no urethane, it probably contains sulfites. Or beer may. And in any case, both are laden with a more pernicious chemical, alcohol. Does this all mean that modern life is burdening Americans with more and more worries? No—just different ones. Worries grow stale and need to be changed. It's the disposition to worry that endures."[1]

The flippancy of the *Times* editorial has been emblematic of the general tone of news coverage of chemical risks in recent years. For example, there was a strong media counterreaction to the Alar scare in 1989. Scientists also began to speak out freely about what they saw as the public's misplaced concerns over synthetic organic chemicals and cancer. In the early 1980s, coalitions of business interests were beginning to form to overturn the Delaney clause of the Food and Drug Amendments of 1958. Among some scientists there was a rising skepticism that animal bioassays utilizing high doses of a chemical could in fact shed light on whether the substance was a human carcinogen. Interest among science journalists shifted from a chemical to a genetic etiology of cancer. The idea of cancer as an inherited disease fit in well with the Human Genome Project initiative and the trend toward the geneticization of illness.

By the mid-1990s, the environmental endocrine hypothesis had begun to invoke new metaphors of chemical risk, both within the scientific community and among the general public.

Within the scientific community the concept of endocrine disrupters provided an overarching framework that connected diverse areas of research. A number of human diseases whose causes were not well understood were beginning to be reexamined in terms of the mechanisms associated with endocrine disrupters. Among them were diseases of the male and female reproductive system, immune system, and thyroid gland, as well as breast and testicular cancers.

The publicity over the hypothesis prompted some scientists to question whether the human or wildlife diseases they were studying were in some way related to chemical exposures at some critical point in the development of the organism. The traditional dichotomy of genes and environment was being recast. During gestation, the fetus is highly sensitive to hormonal changes in the uterine environment. Chemicals trapped in the mother's fatty tissue or blood are mobilized during pregnancy or lactation and then transferred to the embryo or neonate. Chemicals with hormone-disrupting properties can affect brain development or disturb anatomical balance. Scientists devised animal models to study how a one-time chemical exposure in utero at the appropriate stage of embryonic development can result in irreversible abnormalities that may not show up in humans for decades.

According to some interpretations of the somatic mutation theory of cancer, cells whose DNA has been chemically altered may remain quiescent for years until the conditions are right for cell proliferation and tumor growth. Synthetic chemicals that behave like hormones are also believed capable of transforming fetal cells in a manner that eventually expresses disease states or developmental abnormalities.

A recent study of the amniotic fluid of unborn babies lends plausibility to the view that the human fetus is exposed to high enough concentrations of foreign chemicals to affect development. Claude Hughes, a specialist in obstetrics and gynecology at Cedars-Sinai Medical Center in Los Angeles, reported at a meeting of the Endocrine Society on June 14, 1999, that significant concentrations of 1,1-dichloro-2,2-bis ethylene (DDE) were found in the amniotic fluid of 53 women at between 16 and 20 weeks of gestation. The concentrations of DDE (a testosterone antagonist) found in the fluid were at about the same level as natural male hormones found in female fetuses and at about 50 percent of the level of natural male hormones found in male fetuses. These studies indicate that the fetus is exposed to relatively high concentrations of a hormone antagonist during early stages of development.[2]

In the 1996 consensus statement from the work session on "Chemically Induced Alterations in the Developing Immune System" (see Appendix B), 18 scientists expressed certainty that "impairment of the immune system can result from alterations in the development of the immune system and may be long-lasting. The effects may not be manifested at hatching or birth and may not be expressed until the animal or human reaches adulthood."[3] Although the disease end points of cancer are not defined by a single cell type or disease mechanism, the unifying themes in cancer are abnormal cell types, cell proliferation, and cell metastasis. The unifying theme for endocrine disrupters lies not in a particular disease outcome or phenotypic condition but in the role that xenobiotic chemicals play in scrambling messages during and after fetal development. For example, there are polar bears in Norway growing to maturity that were born with both male and female sex organs. The abnormalities of the bears' reproductive organs have been attributed to their exposure to polychlorinated biphenyls (PCBs) in their diet.

In many respects the environmental endocrine hypothesis has already met with successes. The concept of endocrine disrupters has introduced a new hypothesis-generating space that encourages inquiry into the study of chemically induced, hormonally mediated diseases. Scientists who remain agnostic about whether endocrine disrupters are responsible for human abnormalities nonetheless place considerable confidence in laboratory studies and wildlife results demonstrating the endocrine effects of selected chemicals. For more than 30 years the scientific literature has documented numerous cases of abnormal development in wildlife among diverse species. These studies, when examined through the lens of the theory of endocrine disrupters, will both contribute to and be illuminated by that theory.

The publicity surrounding and governmental response to endocrine disrupters in the mid-1990s have linked many small and poorly funded research initiatives under the rubric of a unified hypothesis that has elevated the importance of wildlife studies, comparative endocrinology, and research into hormone receptor mechanisms. As with cancer, there is an emergent endocrine disrupter constituency that is manifest in research funding priorities, citizen advocacy, conferences, and regulation. It has not turned society's attention away from cancer, but it has expanded the vistas of cancer researchers to include investigations into xenobiotic sources for hormone-dependent cancers.[4]

As the scope of the environmental endocrine hypothesis expanded, it also became more vulnerable to contested claims. However, the hypothesis is not structured around a tightly connected set of propositions. It is not a theory in the conventional sense, such as one finds in physics: a set of propositions and rules of correspondence linking theoretical terms to observables. Rather, the hypothesis is a loosely constructed framework—a convex lens built from confirmed results and more speculative hypotheses that directs investigators to focus on certain mechanisms and associations between chemical exposures and human and animal effects. Under its umbrella are more tightly framed, testable conjectural propositions, such as whether in utero exposure to endocrine-disrupting chemicals results in developmental abnormalities or adult diseases. The overall hypothesis can tolerate some inconsistencies, negative or nonconfirmatory evidence, and even the refutation of selected subhypotheses without sustaining a fatal blow. Its resilience lies in the fact that some of the basic scientific claims on which the general hypothesis rests are solidly grounded on empirical results. Moreover, scientists have investigated several biochemical pathways that provide a mechanistic explanation of the laboratory and wildlife phenomena that illustrate the hormone-like effects of environmental contaminants.

Critics who argue that the theory of endocrine disrupters is nothing more than a hypothesis are trivializing the importance of animal data and of the progress science has made in understanding how environmental chemicals can interfere with biochemical pathways at critical stages of an organism's development. As illustrations, the following scientific claims involve replicated results that are largely uncontested within the scientific community:

— Wildlife exposed to high (and in some cases moderate) concentrations of certain synthetic chemicals exhibit responses typically induced by sex steroids.
— In vitro studies demonstrate that synthetic chemicals can bind with and activate hormone receptors, resulting in gene expression.
— Exposure of pregnant mice to very low concentrations of certain synthetic chemicals results in offspring with lower sperm production and increased prostate size.
— Some groups of synthetic organic chemicals can induce human breast cells to proliferate in culture in a manner analogous to the action of the potent female hormone estradiol.
— Male fish and alligators exposed to industrial effluent exhibit signs of feminization—a result that can be replicated in the laboratory when eggs are exposed to certain varieties of synthetic chemicals.
— Persistent organic chemicals build up in human tissue and are passed to the developing fetus during pregnancy and to the neonate during lactation.
— Scientists can impair the immune systems and alter the behavior of laboratory animals by exposing them in utero to low doses of selected synthetic organic chemicals.

Even the most ardent skeptics of the hypothesis do not deny that these and several other correlative propositions are solidly grounded. For this reason, some scientists wish to distinguish those endocrine disrupter hypotheses that have been confirmed from those that are still uncertain. Among the knowledge claims with some evidentiary support that have been contested and for which there is not a broad scientific consensus are the following:

— There has been a steady worldwide decline in the density, motility, and health of human sperm over several decades.
— Organochlorines are a causal agent in human breast cancer.
— Cognitive and behavioral abnormalities in children are a result of their in utero exposure to endocrine disrupters.
— Diseases such as testicular and prostate cancer, hypospadia, undescended testes, and assorted abnormalities of reproductive development can be traced to the exposure of pregnant women to ambient levels of endocrine disrupters.

Danzo voices the general consensus: "A number of experimental studies reinforce the concept that environmental xenobiotics can have detrimental effects on the reproductive health of animals. Circumstantial evidence is accumulating that environmental xenobiotics may be disrupting reproductive processes in human males."[5]

When it comes to determining the cause of a human disease, the ethical limits on science are obvious—it is unethical to do a controlled experiment examining the effects of human exposure to industrial chemicals. Epidemiological results are limited because they cannot yield the causal power of controlled experiments. Thus, a statistically significant association such as that reported in the *New England Journal of Medicine,* between the consumption of PCB-laden fish by pregnant women and reduced intelligence and attention deficit problems in their children, will be subject to the standard caveats that confounding variables unknown to the investigators may be at work.[6]

The best animal studies, on the other hand, will be seen as suggestive but not conclusive in predicting human effects. But even without controlled human experiments, the confidence levels connecting in utero or neonatal exposure to endocrine disrupters with childhood or adult abnormalities can be improved with the convergence of many types of indirect studies. As we saw in Chapter 4, this approach to building a causal story without the benefit of controlled experiments is sometimes referred to as the weight-of-evidence approach. A report by the U.S. Agency for Toxic Substances and Disease Registry of the Department of Health and Human Services states that a necessary and reasonable alternative to causal determinations when establishing policy "may be a critical assessment of the overall 'weight of evidence' of available science to serve as a surrogate of 'causality.' "[7] This view is based on the premise that animals are sentinels for human health events. The agency argues that the weight of evidence across species and phyla can be used as the basis for environmental policy, and it has used the approach to implicate PCBs in the causation of human disease:

> Each of these studies, whether an epidemiologic study, a laboratory study, or the findings of wildlife biologists, could be compared to the lens of a microscope. Like the lens of a microscope, they can vary in terms of their resolving power and quality. They are also focused on different populations at different points in time. Yet there are remarkable parallels in some of the findings that have been repeated, and these findings transcend both geographical and phyletic boundaries. Despite the limits and weaknesses of individual pieces of research, the collective weight of evidence indicates that certain PCB/dioxin-like compounds found in fish in the Great Lakes–St. Lawrence basin (and elsewhere) can cause neurobehavioral deficits.[8]

Following the same line of thinking, the International Agency for Research on Cancer adopted the view that "in the absence of adequate data on humans, it is biologically plausible and prudent to regard agents and mixtures for which there is sufficient evidence of carcinogenicity in experimental animals as if they presented a carcinogenic risk to humans."[9]

The weight-of-evidence approach is used to evaluate specific cause-effect scenarios. As more instances of endocrine disturbances in wildlife linked to environmental toxicants are revealed, the generalized theory that humans are at risk from ambient levels of endocrine disrupters will pick up adherents. Yet, ironically, by placing under a single theoretical framework such diverse effects as sperm decline, cancer, immune suppression, and cognitive dysfunction, a false impression is created that a single mode of action may be at work. Likewise, there is also a false impression that any negative finding weakens the general theory. In fact, the edifice of that theory is being constructed at the foundations, stone by stone, with each additional fragment of evidence.

Critics are correct in pointing to the oversimplification of biochemical explanations, particularly as they appear in the popular media and advocacy publications. The burgeoning interest in endocrine disrupters has meant that studies once believed quite distinct and idiosyncratic are being reexamined through a single lens. Not surprisingly, first-generation efforts to account for the data are giving way to new levels of complexity. The same chemical may behave under some circumstances like an estrogen and under other circumstances like an antiestrogen. Chemicals that disrupt cognitive or immune function may do so by modes of action that do not involve hormone receptors. As Gillesby and Zacharewski note, "the mechanisms of action of these estrogen-like substances are extremely diverse and may encompass several different signal pathways. . . . The diverse mechanisms of action and chemical structures make the study of endocrine disrupters an extremely challenging and complex process."[10]

The fact that a particular chemical or a particular species fails to corroborate one mechanism of hormone disruption does not invalidate the utility of the general framework but instead suggests that, within its scope, the mechanistic variations of signaling disruptions by various xenobiotics are yet to be worked out.

We do not need confirmed instances of human disease or actual body counts to appreciate some of the successes of the environmental endocrine hypothesis. It has fostered new crossdisciplinary collaborations that have been responsible for more integrative approaches to the investigation of chemical hazards. It has rekindled among scientists and scientific funding agencies in governments worldwide the idea that environmental factors play a role in the abnormal development of species—particularly chemicals stored in the body during pregnancy and lactation or those that accumulate in the environment that can affect the developing fetus. As an example, Japan's leading environmental agency has established the world's first scientific society dedicated to the study of endocrine disrupters, in response to sustained coverage of the issue on the front pages of the country's newspapers. The hypothesis has brought forward new inquiring minds from many disciplines to examine diseases of unknown origin, such as prostate cancer and hypospadia, whose incidence rates have been rising in industrialized countries.

At the same time, the study of the endocrine effects of chemicals has begun to focus attention on the limits of traditional toxicology. Chemical imprinting of the brain may not leave behind the standard biomarkers of chemical toxicity, such as cell death, chromosomal damage, and tumors. New methods for studying the delayed effects of chemicals grow out of the diethylstilbestrol (DES) experience and may apply as well to other hormone mimics or antagonists.

The complexity of sorting out causality when faced with multiple chemical exposures over a lifetime has also brought new awareness of the precautionary principle, and the environmental endocrine hypothesis has provided a forum within which to explore the application of the principle to public policy.

For all these reasons, the term *endocrine disrupter* has been permanently etched into the scientific lexicon and the social agenda. Henceforth, when scientists are asked to inquire into whether chemicals are biologically active, endocrine disruption will take its proper place beside acute toxicity, carcinogenicity, mutagenicity, genotoxicity, and teratogenicity. This is a legacy of no small proportions.

Epilogue

On August 4, 1999, the National Research Council of the U.S. National Academy of Sciences made public its long-awaited report on endocrine disrupters, titled *Hormonally Active Agents in the Environment.* Prepared by a 16-member panel of scientists (one member fewer than the original committee) that had begun its inquiry in 1995, the 414-page report contains three important messages. *First,* the committee was in agreement on the seriousness of the scientific problem of endocrine-disrupting chemicals (which it refers to as Hormonally Active Agents, or HAAs), and it concluded that the issue should be pursued through a broad range of research, monitoring and testing of wildlife and human populations, and the development of assays for screening chemicals. The NAS findings validate the work of scientists who have supported and advanced the environmental endocrine hypothesis as well as the actions taken by the United States and other nations to study the health and environmental effects of endocrine disrupters. The committee was not in agreement on whether HAAs in the environment present a major health threat and require an immediate and coordinated response.

Areas in which the report is consonant with prior claims made on behalf of the environmental endocrine hypothesis are illustrated by such findings as:

— "Adverse reproductive and developmental effects have been observed in human populations, wildlife, and laboratory animals as a consequence of exposures to HAAs"(p. 3);
— prenatal exposure to certain chemicals can cause lower birth weight and shorter gestation and has been correlated with deficits in memory and IQ and delayed neuromuscular development (p. 3);
— laboratory studies involving a variety of animals have shown that in utero exposures of these animals to different concentrations of certain HAAs can produce structural and functional abnormalities of the reproductive tract (p. 3);
— health effects discovered in wildlife are also observed in laboratory animals exposed to the same chemicals to which the wildlife are exposed (p. 4).

The committee also acknowledged that "biologic responses to some HAAs might be greater at low doses than at high doses" (p. 8), signifying that the chemicals do not necessarily exhibit a monotonic dose-response.

Second, the report identifies gaps in our knowledge which limit what can be inferred about the effects of chemicals. It is tentative about which chemical causes which effect, and through which endocrine-disrupting mechanism, although it cites a growing body of circumstantial and probabilistic evidence linking synthetic chemicals that have endocrine-disrupting properties to developmental and reproductive abnormalities in wildlife. With respect to possible effects of endocrine disrupters on humans, the report also points to significant data gaps in critical areas of concern: "Exposure to HAAs and their possible effects during susceptible periods such as fetal life or pregnancy and transgenerational effects have not been evaluated in human studies" (p. 257).

Some of the strongest language is found in the report's conclusions on wildlife effects. "Environmental HAAs probably have contributed to declines in some wildlife populations, including fish and birds of the Great Lakes and juvenile alligators of Lake Apopka, and possibly to diseases and deformities in mink in the United States, river otters in Europe, and marine mammals in European waters" (p. 6). "There is evidence that certain synthetic, persistent, bioaccumulative hydrocarbons have caused effects on wildlife reproduction. . . ." (p. 279).

Third, the report cites disagreement among committee scientists on the interpretation of and generalization from some studies. The statement that best captures the tenor of the study appears in the first paragraph of the executive summary: "Although it is clear that exposures to HAAs at high concentrations can affect wildlife and human health, the extent of harm caused by exposure to these compounds in concentrations that are common in the environment is debated" (p. 1). Hundreds of scientific studies formed the basis of the council's report. Many statements of interpretation of trends in the data cited in the report for specific human or animal effects were balanced with a qualifying phrase. The qualifiers included: lack of cause and effect relationships; paucity of human exposure data; failure to understand the underlying mechanism of chemical action on biological systems; limits to extrapolation from results in laboratory animals to effects in humans; and the difficulty of sorting out hormonally active synthetic chemicals from HAAs found in nature.

The executive summary and introduction of the study discuss the divergent viewpoints held among members of the committee. "It became clear as the work of the committee progressed that limitations and uncertainties in the data could lead to different judgments among committee members with regard to interpreting the general hypothesis, determining appropriate sources of information, evaluating the evidence, defining the agents of concern, and evaluating environmental and biological variables." The differences among committee members were only in part the result of gaps in the science. "Some differences appear to stem from different views of the value of different kinds of evidence obtained by experiments, observations, weight-of-evidence approaches, and extrapolation of results from one compound or organism to others, as well as allowable sources of information and criteria for arriving at meaningful conclusions and recommendations" (p. 2). "Much of the division among committee members appears to stem from different views of how we come to know what we know. How we understand the natural world and how we decide among conflicting hypotheses about the natural world is the province of epistemology. Committee members seemed to differ on some basic epistemologic issues, which led to different interpretations and conclusions on the issues of HAAs in the environment" (p. 13).

A decision process that produces a consensus report by a divided panel on the state of scientific evidence for an issue that was born out of controversy must inevitably result in understatement and caution in the interpretation of the science. Any statement in the report indicating support for the general hypothesis had to pass through the filter of skeptics and naysayers both on the committee and in external review panels. This may explain why the report contained so many qualifying phrases, gave few definitive explanations of endocrine-disrupter etiology, and took a very conservative view of causality. Finally, having been subjected to critical evaluation by some of its harshest skeptics, the environmental endocrine hypothesis has emerged from its review by the National Research Council as a legitimate and promising approach for the study of industrial and agricultural chemicals.

APPENDIX

Chronology of Key Events in the Development of the Environmental Endocrine Hypothesis

1938 Diethylstilbestrol (DES) reported as a new synthetic estrogen.

1941 U.S. Food and Drug Administration approves the use of DES for hormone therapy and as a relief for menopausal disorders.

1947 U.S. Department of Agriculture approves the use of DES to promote the growth of chickens.

1949 U.S. Food and Drug Administration approves the use of DES for the prevention of miscarriages.

1954 U.S. Department of Agriculture extends its approval of DES to cattle and sheep through implants and in feed.

1959 DES is found to produce cancer in experimental animals. Its sale and use on chickens are suspended.

1962 Publication of *Silent Spring* by Rachel Carson.

1971 Association is found between mothers who took DES and a rare form of vaginal cancer in their daughters. U.S. Food and Drug Administration bans the use of DES on animals.

1972 U.S. Environmental Protection Agency (EPA) bans the pesticide dichloro-diphenyl trichloroethane (DDT). U.S. Food and Drug Administration warns physicians against prescribing DES for pregnant women.

1977 EPA bans the manufacture and use of polychlorinated biphenyls (PCBs).

1979 (September 10–12). Conference, "Estrogens in the Environment," sponsored by the National Institute of Environmental Health Sciences, Raleigh, North Carolina. Published proceedings: McLachlan (1980).

1981 (November 1–4). Eleventh Banbury Conference, "Environmental Factors in Human Growth and Development," Cold Spring Harbor Laboratory. Published proceedings: Hunt et al. (1982).

1985 (April 10–12). Conference, "Estrogens in the Environment II: Influences on Development," sponsored by the National Institute of Environmental Health Sciences, Raleigh, North Carolina. Published proceedings: McLachlan (1985).

1990 Publication of studies for the International Joint Commission on wildlife effects of chemicals around the Great Lakes. See Colborn et al. (1990).

1991 (July 26–28). Wingspread Work Session, "Chemically Induced Alterations in Sexual and Functional Development: The Wildlife/Human Connection," Racine, Wisconsin, convened by T. Colborn. Published proceedings: Colborn et al. (1992).

1991 (September 30–October 4). International Workshop on the Impact of the Environment on Reproductive Health, sponsored by the World Health Organization and other groups. Published proceedings: Skakkebaek et al. (1993).

1991 (October 2). U.S. Senate, Committee on Governmental Affairs, hearings on "Government Regulation of Reproductive Hazards." Published proceedings: U.S. Congress, Senate, Committee on Governmental Affairs (1992).

1992 E-SCREEN assay for xenoestrogens developed by Ana Soto and Carlos Sonnenschein. See Soto et al. (1992, 1995).

1993 (October 21). U.S. House of Representatives, Committee on Energy and Commerce, Subcommittee on Health and the Environment, hearings on "Health Effects of Estrogenic Pesticides." Published proceedings: U.S. Congress, House, Committee on Energy and Commerce, Subcommittee on Health and the Environment (1994).

1993 (December 10–12). Wingspread Work Session, "Environmentally Induced Alterations in Development: A Focus on Wildlife," Racine, Wisconsin, convened by T. Colborn. See Bantle et al. (1995).

1994 (January 9–11). Conference, "Estrogens in the Environment III: Global Health Implications," sponsored by the National Institute of Environmental Health Sciences, Raleigh, North Carolina, organized by J. A. McLachlan and K. S. Korach. Published proceedings: McLachlan and Korach (1995).

1994 (September 4–5). BBC documentary *Assault on the Male* airs on the Discovery Channel.

1995 (January 23–27). Workshop, "Male Reproductive Health and Environmental Chemicals with Estrogenic Effects," convened at Rigshospitalet, Copenhagen, at the request of the Danish Environmental Protection Agency and the Danish Ministry of Environment and Energy. Report: Danish Environmental Protection Agency (1995).

1995 (February 10–12). Wingspread Work Session, "Chemically Induced Alterations in the Developing Immune System: The Wildlife/Human Connection," Racine, Wisconsin, convened by T. Colborn. Consensus statement: Barnett et al. (1996).

1995 (April 10–13). Office of Research and Development–EPA workshop on endocrine disrupters, Research Triangle Park, Raleigh, North Carolina, convened by R. J. Kavlock. Published proceedings: Kavlock et al. (1996).

1995 (June 9). Fourth North Sea Conference, Esbjerg, Denmark. International meeting of environment ministers of countries bordering the North Sea; agreement reached on phase-out of persistent organic pollutants.

1995 (July 21–23). Wingspread Work Session, "Chemically Induced Alterations in the Functional Development and Reproduction of Fishes," Racine, Wisconsin, convened by T. Colborn. Published proceedings: Rolland et al. (1997).

1995 (October 24–26). First meeting of the National Academy of Sciences–National Research Council study panel on Hormone-Related Toxicants in the Environment, Washington, D.C.

1995 (October 29–November 1). Conference, "Developmental Neurotoxicity of Endocrine Disruptors: Dioxin, PCBs, Pesticides, Metals, Psychoactive and Therapeutic Drugs," Hot Springs, Arkansas.

1995 (November 5–10). Erice Work Session, "Environmental Endocrine-Disrupting Chemicals: Neural, Endocrine, and Behavioral Effects," Erice, Sicily, Italy, convened by T. Colborn. Consensus statement: Brouwer et al. (1998).

1996 (March). Publication of *Our Stolen Future,* by Theo Colborn, Dianne Dumanoski, and John Peterson Myers.

1996 (May 15–16). EPA-sponsored workshop on screening for chemicals suspected of disrupting hormonal systems, Washington, D.C.

1996 (May 26–30). Thirty-Ninth Annual Conference on Great Lakes Research, focused on endocrine disrupters, University of Toronto.

1996 (August). Food Quality Protection Act passed; requires the EPA to obtain data about the potential hormone-disrupting effects of pesticide residues in food.

1996 (August). Safe Drinking Water Estrogenic Substances Screening Program Act amends the Safe Drinking Water Act (42 USC 300f et seq.); requires the EPA to establish a screening program for estrogenic substances.

1996 (September 27–29). Wingspread Work Session, "Health Effects of Contemporary-Use Pesticides: The Wildlife/Human Connection," Racine, Wisconsin, convened by T. Colborn. Consensus statement: Brock et al. (1999).

1996 (October 16). EPA charters the Endocrine Disrupter Screening and Testing Advisory Committee (EDSTAC).

1996 (December 2–4). European Workshop on the Impact of Endocrine Disrupters on Human Health and Wildlife, Weybridge, England. Published proceedings: International Organization for Economic Cooperation and Development (1997).

1997 (April 10–13). International Organization for Economic Cooperation and Development–Society for Environmental Toxicology and Chemistry–European Community Workshop on Endocrine Modulators and Wildlife Assessment and Testing, Amsterdam.

1997 (July 14). The Great Lakes Endocrine Disruptor Symposium, Chicago, Illinois.

1997 (July 20–23). Conference, "Estrogens in the Environment IV: Linking Fundamental Knowledge, Risk Assessment, and Public Policy," sponsored by the National Institute of Environmental Health Sciences and the National Toxicology Program, Arlington, Virginia.

1998 (June 2). *Fooling with Nature,* a documentary on endocrine-disrupting chemicals, airs on the Public Broadcasting System's *Frontline.*

1998 (July 12–17). Gordon Research Conference on Environmental Endocrine Disrupters, Plymouth State College, Plymouth, New Hampshire.

1998 (August). Final Report, Endocrine Disrupter Screening and Testing Advisory Committee, Office of Prevention, Pesticides and Toxic Substances, U.S. Environmental Protection Agency, Washington, D.C.

1998 (December 11–13). International Conference on Endocrine Disrupters, Kyoto, Japan.

1999 (July). National Research Council. *Hormonally Active Agents in the Environment.* Washington, D.C.: National Academy Press.

APPENDIX

B

Consensus Statements from Endocrine Disrupter Work Sessions, 1991–1996

Title of work session	*Place/date of meeting*	*Publication of consensus statement*	*N**
Chemically Induced Alterations in Sexual and Functional Development: The Wildlife/Human Connection	Wingspread Conference Center Racine, Wisconsin July 26–28, 1991	Bern et al. (1992)	21
Environmentally Induced Alterations in Development: A Focus on Wildlife	Wingspread Conference Center Racine, Wisconsin December 10–12, 1993	Bantle et al. (1995)	23
Chemically Induced Alterations in the Developing Immune System: The Wildlife/Human Connection	Wingspread Conference Center Racine, Wisconsin February 10–12, 1995	Barnett et al. (1996)	18
Chemically Induced Alterations in the Functional Development and Reproduction of Fishes	Wingspread Conference Center Racine, Wisconsin July 21–23, 1995	Benson et al. (1997)	22

Title of work session	*Place/date of meeting*	*Publication of consensus statement*	*N**
Environmental Endocrine-Disrupting Chemicals: Neural, Endocrine, and Behavioral Effects	Ettore Majorana Centre for Scientific Culture Erice, Sicily, Italy November 5–10, 1995	Brouwer et al. (1998)	18
Health Effects of Contemporary-Use Pesticides: The Wildlife/Human Connection	Wingspread Conference Center Racine, Wisconsin September 27–29, 1996	Brock et al. (1999)	23

*N = Number of scientist signatories.

APPENDIX

C

Reviews of *Our Stolen Future*

Author	*Profession*	*Publication*	*Date*	*Type of review*	*Rating*
Moghissi	Scientist	*Environment International*	1996	Book review	Negative
Foulkes	Doctor	*Fluoride*	1996	Book review	Positive
Meadows	Scientist	*Los Angeles Times*	1-31-96	Op-ed article	Positive
Waldholz	Staff writer	*Wall Street Journal*	3-7-96	Book review/news	Mixed
Carpenter	Staff writer	*U.S. News & World Report*	3-11-96	Book review/news	Positive
Mathews	Policy adviser	*Washington Post*	3-11-96	Op-ed article	Positive
Malkin	Policy adviser	*Seattle Times*	3-12-96	Op-ed article	Negative
Editorial	—	*Washington Times*	3-13-96	Editorial	Negative
Begley and Glick	Staff writers	*Newsweek*	3-21-96	Book review/news	Mixed
Raeburn	Staff writer	*Business Week*	3-18-96	Book review/news	Mixed
Kolata	Staff writer	*New York Times*	3-19-96	Book review/news	Negative
Dansereau	Policy adviser	*Seattle Times*	3-20-96	Op-ed article	Positive
Snow	Staff writer	*Hattiesburg American*	3-21-96	Book review	Mixed
Zimmerman	Scientist	*Philadelphia Inquirer*	3-24-96	Book review	Positive

Author	Profession	Publication	Date	Type of review	Rating
Bailey	Television producer	*Washington Post*	3-31-96	Op-ed article	Negative
Editorial	—	*Richmond Times-Dispatch*	3-31-96	Book review	Negative
Fumento	Science writer	*Sacramento Bee*	3-31-96	Op-ed article	Negative
Weiss and Lee	Staff writers	*Washington Post*	3-31-96	Book review/news	Mixed
Cortese	Scientist	*Environmental Science & Technology*	4-1-96	Book review	Positive
Johnson	Staff writer	*Environmental Science & Technology*	4-1-96	Book review/news	Mixed
Lucier and Hook	Scientists	*Environmental Health Perspectives*	4-1-96	Editorial	Mixed
Moomaw	Scientist	*Chemical & Engineering News*	4-1-96	Book review	Positive
Sullivan	Scientist	*Los Angeles Times*	4-1-96	Op-ed article	Negative
Taylor	Staff writer	*Seattle Post-Intelligencer*	4-2-96	Book review/news	Mixed
Hertsgaard	Author	*New York Times Book Review*	4-7-96	Book review	Positive
Demarest	Scientist	*San Francisco Chronicle*	4-14-96	Book review	Positive
Lee	Staff writer	*Washington Post*	4-14-96	Book review	Positive
Springston	Staff writer	*Richmond Times-Dispatch*	4-14-96	Book review/news	Mixed
Carey	Senior correspondent	*Business Week*	4-18-96	Book review	Negative
Windsor	Staff writer	*New American*	4-29-96	Book review	Negative
Bonner	Science writer	*New Scientist*	5-4-96	Book review	Positive
Johnson	Policy adviser	*San Francisco Examiner*	5-5-96	Op-ed article	Positive
Beatty	Editor	*Audubon*	5-6-96	Book review	Positive
Hirshfield et al.	Scientists	*Science*	6-7-96	Book review	Positive
Scialli	Doctor	*Reproductive Toxicology*	6-18-96	Book review	Negative
Zeeman	Scientist	*BioScience*	7-96	Book review	Positive
Baden and Noonan	Foundation chairman and research assistant	*American Enterprise*	7-8-96	Book review	Negative

Author	*Profession*	*Publication*	*Date*	*Type of review*	*Rating*
Kamrin	Scientist	*Scientific American*	9-96	Book review	Negative
Giddens	Staff writer	*London Review of Books*	9-5-96	Book review	Mixed
Lemonick	Staff writer	*Time*	9-19-96	Book review/news	Mixed

NOTES

Chapter 1: Scientific Developments

1. Burlington and Lindeman 1950:51.
2. Carson 1962:197.
3. Carson 1962:209.
4. Carson 1962:209–10.
5. Carson 1962:210.
6. Carson 1962:211.
7. Burlington and Lindeman 1950.
8. McLachlan and Newbold 1987:25.
9. Palmlund 1996.
10. Apfel and Fisher 1984:14.
11. Cottrell 1971.
12. Palmlund 1996.
13. Stillman 1982.
14. Herbst et al. 1971.
15. Herbst and Scully 1970; Herbst et al. 1971; Stillman 1982.
16. Herbst and Bern 1981:156, 197.
17. Herbst et al. 1971.
18. McLachlan 1995.
19. McLachlan 1980.
20. Newbold et al. 1990.
21. McLachlan and Newbold 1987:25.
22. McLachlan 1995.
23. Duax et al. 1984.
24. Cadbury 1997:77.
25. Myers 1979.
26. Gellert 1978; Hammond et al. 1979.
27. Colborn 1981.
28. Colborn 1985.
29. Colborn et al. 1990:182.
30. Colborn 1995.
31. Fox 1996; Gilbertson 1996.
32. Colborn et al. 1996:25.
33. Colborn et al. 1996:26.
34. Colborn et al. 1990:139.
35. Wapner 1995.
36. Colborn 1995.
37. Colborn et al. 1990.
38. Myers 1998.
39. Colborn 1991:109.
40. Vom Saal and Bronson 1980; vom Saal et al. 1983.
41. Vom Saal 1996.
42. Ibid.
43. Myers 1999.

44. Vom Saal 1996.
45. McLachlan 1980:vi–vii.
46. McLachlan and Newbold 1987:25.
47. McLachlan and Newbold 1987:26.
48. McLachlan 1995.
49. Cited in Wapner 1995:21.
50. Colborn and Clement 1992.
51. Myers 1999.
52. Houghton et al. 1990:xi–xii.
53. Colborn and Clement 1992.
54. Skakkebaek 1996.
55. Skakkebaek et al. 1993.
56. Skakkebaek 1996.
57. Carlsen et al. 1992:612.
58. Suominen and Vierula 1993.
59. Sharpe and Skakkebaek 1993.
60. Ibid.
61. Skakkebaek 1996.
62. Whorton et al. 1979; Whorton and Milby 1980.
63. Thrupp 1991.
64. Bibbo et al. 1977.
65. Sharpe and Skakkebaek 1993: 1392. The "sea of estrogens" metaphor is attributed by Sharpe and Skakkebaek to Field et al. 1990.
66. Sharpe and Skakkebaek 1993:1393.
67. Wright 1996:44.
68. Danish Environmental Protection Agency 1995.
69. Wright 1996.
70. Fisch and Goluboff 1996.
71. Paulsen et al. 1996.
72. Kolata 1996c.
73. Toppari et al. 1996.
74. Danish Environmental Protection Agency 1995.
75. Toppari et al. 1996:768.
76. Swan et al. 1997:131.
77. Anonymous 1994.
78. Erikson 1995:1508.
79. Gorbach et al. 1984.
80. Stoll 1969:36.
81. Gorbach et al. 1984:39.
82. Lahita et al. 1981; Schneider et al. 1982.
83. Osborne et al. 1993.
84. Davis et al. 1993.
85. Davis et al. 1993:372.
86. Davis and Bradlow 1995.
87. Davis et al. 1997.
88. Raloff 1993.
89. Clorfene-Casten 1993:53.
90. Colborn et al. 1996:184–85.
91. Meilahn et al. 1998; Telang et al. 1998.
92. Safe and McDougal 1997:7.
93. Bradlow 1997.
94. Hunter et al. 1997.
95. Safe 1997b:1304.
96. Tye 1998.
97. Steingraber 1997a:267.
98. Daly et al. 1989.
99. Dumanoski pers. comm. 1998; the Hauser paper is Hauser et al. 1993.
100. Ibid.
101. Alleva et al. 1998.
102. Needleman and Gatsonis 1990.
103. Rogan et al. 1988.
104. Ibid.; Chen et al. 1992, 1994.
105. Jacobson et al. 1985.
106. Ibid.
107. Gladen et al. 1988.
108. Jacobson et al. 1990b.
109. Daly et al. 1989; Daly 1991, 1993.
110. Alleva et al. 1998:2.
111. B. Weiss, http://www.envirotrust. com/stolesupp.html.
112. Jacobson and Jacobson 1996.
113. Brody 1996.
114. Fox 1992; Reijnders and Brasseur 1992.
115. Colborn et al. 1993.
116. Danish Environmental Protection Agency 1995.
117. Colborn et al. 1993; Davis et al. 1993; Safe 1995a.
118. Davis and Bradlow 1995.
119. Reijnders and Brasseur 1992.
120. Danish Environmental Protection Agency 1995.

Chapter 2: The Emergence of a Public Hypothesis

1. Krimsky and Plough 1988.
2. Colborn et al. 1993; Safe 1995a.
3. Krimsky and Golding 1992.
4. U.S. Congress, Senate 1992.
5. U.S. Congress, House 1994.

6. U.S. Congress, Senate 1992:2.
7. U.S. General Accounting Office 1992:4.
8. Colborn 1991:95.
9. U.S. Congress, Senate 1992:51.
10. Jacobson et al. 1985, 1990b.
11. U.S. Congress, Senate 1992:54.
12. U.S. Congress, Joint Committee 1994:6.
13. U.S. Congress, House 1994:2.
14. U.S. Congress, House 1994:72.
15. U.S. Congress, House 1994:38.
16. U.S. Congress, House 1994:17, 41, 124.
17. Kaiser 1996.
18. U.S. Congress, Health and Environment 1996:83, 18.
19. Barish 1998.
20. U.S. Congress, Joint Committee 1994:11–12.
21. U.S. Congress, Senate 1995:S17749.
22. U.S. Congress, Health and Environment 1996:148.
23. U.S. Congress, Agriculture 1996:56.
24. U.S. Congress, Agriculture 1996:123.
25. U.S. Congress, Commerce 1996:87.
26. Endocrine Disrupter Screening and Testing Advisory Committee 1998: ES-1.
27. Dumanoski 1996/97:40.
28. Dumanoski 1996/97:43.
29. Carson 1962:14.
30. Dumanoski pers. comm. 1998.
31. Lear 1997:430.
32. Colborn et al. 1996:vi.
33. Anonymous 1996b.
34. McLachlan and Newbold 1987:25.
35. Newbold and McLachlan 1996.
36. Kavlock et al. 1996:715.
37. National Academy of Sciences 1994.
38. Environmental Protection Agency 1997.
39. Environmental Protection Agency 1997:2.
40. International Organization for Economic Cooperation and Development 1997:8–9.
41. Silent Spring Institute 1996:II:3.
42. Soto et al. 1995.
43. Greenpeace International 1995.
44. Silverstein 1996:28.
45. Whelan et al. 1996.
46. Environmental Media Services 1996.
47. Anonymous 1996c.
48. Luoma 1992a.
49. Luoma 1992b.
50. Peterson 1993.
51. Sharpe and Skakkebaek 1993.
52. Healy 1993.
53. Beil 1993.
54. Stevens 1994.
55. Cone 1994a–c.
56. Barnard 1994.
57. Raloff 1994a.
58. Raloff 1994b.
59. Weiss 1994.
60. Kay 1994.
61. National Wildlife Federation 1994b.
62. Hiltbrand 1994.
63. Goodman 1994.
64. *Arizona Republic* 1995.
65. *Atlanta Journal and Constitution* 1995.
66. Haybron 1995.
67. Pinchbeck 1996.
68. Wright 1996.
69. Colborn et al. 1996.
70. Carpenter 1996.
71. Crumley et al. 1996.
72. Kolata 1996a,b.
73. Kolata 1996c.
74. Bailey 1996; Weiss and Lee 1996.
75. McKenna 1994.
76. Arnold et al. 1996.
77. Frith 1996; Roberts 1996.
78. Golder 1996.
79. Wapner 1995:21.
80. Fantle 1994.
81. M. Fox 1996.
82. Burger 1996.
83. Toppari et al. 1996:741.
84. Worldwatch Institute 1994:130.
85. Fairley et al. 1996:30.

Chapter 3: Uncertainty, Values, and Scientific Responsibility

1. Planck 1949:33–34.
2. Kuhn 1962:159.
3. Monmaney 1993.
4. Hentschel et al. 1993.
5. Colborn et al. 1993:379.
6. Ashby et al. 1997a:165.
7. Lamb 1997:32.
8. Patlak 1996:542.
9. Ashby et al. 1997a:165.
10. Hammond et al. 1979.
11. Rudolph 1999.
12. Sheehan and vom Saal 1997:36.
13. Sheehan and vom Saal 1997:38.
14. Environmental Protection Agency 1997:2.
15. Safe and Ramamoorthy 1998:22.
16. Wolff and Landrigan 1994:525–26.
17. Steingraber 1997b:686.
18. Altenburger et al. 1996:1157.
19. Safe 1995a.
20. Safe 1995a, 1997a.
21. Hunter et al. 1997.
22. Colborn et al. 1996:180.
23. Environmental Protection Agency 1997:6, 9.
24. Colborn et al. 1996:196, 186.
25. Meadows 1996.
26. Mathews 1996.
27. Anonymous 1996a.
28. J. Johnson 1996; Waldholz 1996.
29. Hileman 1996:28.
30. See, e.g., Malkin 1996; *Washington Times* 1996.
31. Lee 1996.
32. Weiss and Lee 1996.
33. Slovic as quoted in Weiss and Lee 1996:A14.
34. Kolata 1996a,b.
35. Kolata 1996c.
36. Hertsgaard 1996:25.
37. Sullivan 1996:B5.
38. Malkin 1996:B4.
39. Kolata 1996a:C10.
40. Dowie 1998:18, 19.
41. Bailey 1996:C3.
42. Hileman 1996:28.
43. H. D. Johnson 1996.
44. Lucier and Hook 1996:350.
45. Cortese 1996:213A.
46. Zeeman 1996:544.
47. Colborn et al. 1996:vi.
48. Graham 1970:57.
49. Hynes 1989:41–42.
50. Kamrin 1996:178.
51. Hirshfield et al. 1996:1444–45.
52. Krimsky 1982.
53. Funtowicz and Ravetz 1992.
54. Efron 1984:258.
55. Roush 1995.
56. Andreopolis 1980.
57. Krimsky 1982.
58. Zilinskas and Zimmerman 1986; Piller and Yamamoto 1988.
59. *New York Times* 1996.
60. Leary 1998:A13.
61. Kamrin 1996:178.
62. Alleva et al. 1998:5.
63. Vom Saal 1996.
64. Skakkebaek 1996.
65. Safe 1995a.
66. Safe 1997a,b; Safe and McDougal 1997.
67. Cadbury 1997:175.
68. Safe 1997c.
69. Whelan et al. 1996:1.
70. Safe 1994, 1995a.
71. Popper 1959:33.
72. Safe 1995b:785.
73. Hertsgaard 1996:25.
74. Colborn et al. 1996:208.
75. Steingraber 1997a:17.
76. Arnold et al. 1996:1490.
77. Arnold et al. 1996.
78. Chemical Industry Institute of Toxicology 1997.
79. Ashby et al. 1997b; Ramamoorthy et al. 1997.
80. McLachlan 1997:462–63.
81. Weiser et al. 1997:20–21.
82. Fagin and Lavelle 1996.
83. Weiss 1997.
84. Safe 1997a:A14.
85. Katz 1997:A14.
86. Hook and Lucier 1997:784.
87. McKinney 1997:896.
88. Foster 1997:1; the reference is to Sharpe and Skakkebaek 1993.
89. Dodds and Lawson 1936, 1938.
90. Vom Saal et al. 1998:254.

91. Vom Saal 1997.
92. Vom Saal and Welshons 1997.
93. Macilwain 1998.
94. Zinberg 1997:411.
95. Anonymous 1996c.
96. Broton et al. 1995.

Chapter 4: The Policy Conundrum

1. Pauling 1958.
2. Commoner 1971:112.
3. Krimsky 1982.
4. Roan 1989.
5. Gore 1992:318–19.
6. Rosen 1990.
7. Natural Resources Defense Council 1989.
8. Wilson 1940.
9. Glas 1989:137.
10. Ames et al. 1987.
11. Krimsky 1992:19–20.
12. Osborn 1948:61.
13. Carson 1962.
14. Unger et al. 1984.
15. Falk et al. 1992.
16. Wolff et al. 1993.
17. Wolff et al. 1993:652.
18. Krieger et al. 1994.
19. Davis and Bradlow 1995.
20. International Agency for Research on Cancer 1991.
21. Key and Reeves 1994.
22. Quoted in Cadbury 1997:199.
23. Adami et al. 1995.
24. Rivero-Rodriguez et al. 1997.
25. Allen et al. 1997:681.
26. Hunter et al. 1997:1253.
27. Safe 1997b:1303–4.
28. Høyer et al. 1998.
29. Fagin and Lavelle 1996:228.
30. Cranor 1993:118.
31. Huber 1991.
32. Ehrlich and Ehrlich 1996: 199.
33. International Organization for Economic Cooperation and Development 1997:6.
34. Sever et al. 1997.
35. Ginsberg 1996:1501.
36. Abraham and Frawley 1997.
37. Johnstone 1997:5.
38. Cranor 1993:115–16.
39. Krimsky et al. 1991:284; Krimsky et al. 1996.
40. Krimsky and Rothenberg 1998.
41. Cameron and Aboucher 1991:5.
42. Cameron and Aboucher 1991:2.
43. Perrings 1991:166.
44. Cameron and Aboucher 1991:6.
45. Environmental Media Services 1996.
46. Dopyera 1994:A14.
47. Churchman 1947; Shrader-Frechette 1991, 1994; Cranor 1993.
48. Shrader-Frechette 1991.
49. Quoted in Shrader-Frechette 1991:133.
50. Ashford 1997.
51. Stone 1994:308.
52. National Wildlife Federation 1994a,b.
53. Guillette 1994:39–41.
54. National Wildlife Federation 1994a:14.
55. Stevens 1994:C6.
56. U.S. General Accounting Office 1990; Gibbons 1991:25.
57. Endocrine Disrupter Screening and Testing Advisory Committee 1998.
58. Lewis 1990:199.
59. Lave and Upton 1987:283.
60. Fagin and Lavelle 1996:13.
61. Efron 1984:392.
62. Breyer 1993:19.
63. Portney 1992:137.
64. Hynes 1989:102.
65. Lave and Upton 1987:282.
66. Yang 1996:1037.
67. Ashford 1994.
68. Environmental Defense Fund and Boyle 1980:131.
69. Langley 1874:40.
70. Lave and Upton 1987:282.
71. Wiles 1994:33.
72. Dorfman 1982:17.
73. Cushman 1994.
74. Wargo 1996:162.
75. Cohrssen and Covello 1989:85.
76. Lave and Upton 1987:281.
77. U.S. Congress, House, Commerce 1996:87.

78. Environmental Protection Agency, Office of Prevention, Pesticides and Toxic Substances 1997:3.
79. Goldman 1997.
80. Howard 1997:193.
81. Kavlock et al. 1996.
82. Endocrine Disrupter Screening and Testing Advisory Committee 1998:3–2.
83. Ibid.
84. Endocrine Disrupter Screening and Testing Advisory Committee 1998:3–4.
85. Swan et al. 1998.
86. Schettler 1998.
87. National Research Council 1983:3.
88. Endocrine Disrupter Screening and Testing Advisory Committee 1998:532.

Chapter 5: Conclusion: Expanding Paradigms of Chemical Hazards

1. *New York Times* 1988:A26.
2. Van 1999.
3. Barnett et al. 1996:807.
4. Gillesby and Zacharewski 1998.
5. Danzo 1997:294.
6. Jacobson and Jacobson 1996.
7. U.S. Agency for Toxic Substances and Disease Registry 1998:5.
8. U.S. Agency for Toxic Substances and Disease Registry 1998:36.
9. Fung et al. 1995:680.
10. Gillesby and Zacharewski 1998:4.

REFERENCES

Abraham, E. J., and L. S. Frawley. 1997. Octylphenol (OP), an environmental estrogen, stimulates prolactin (PRL) gene expression. *Life Sciences* 60:457–65.

Adami, H. O., L. Lipworth, L. Titusernstoff, et al. 1995. Organochlorine and estrogen related cancers in women. *Cancer Causes and Control* 6:551–66.

Allen, R. H., M. Gottlieb, E. Clute, et al. 1997. Breast cancer and pesticides in Hawaii: the need for further study. *Environmental Health Perspectives* 105(suppl. 3):679–83.

Alleva, E., J. Brock, A. Brouwer, et al. 1998. Statement from the Work Session on Environmental Endocrine-Disrupting Chemicals: neural, endocrine, and behavioral effects. *Toxicology and Industrial Health* 14:1–8.

Altenburger, R., W. Boedeker, M. Faust, and L. H. Grimme. 1996. Regulations for combined effects of pollutants: consequences from risk assessment in aquatic toxicology. *Food and Chemical Toxicology* 34:1155–57.

Ames, B. N., R. Magaw, and L. S. Gold. 1987. Ranking possible carcinogenic hazards. *Science* 236:271–80.

Andreopolis, S. 1980. Sounding board: gene cloning by press conference. *New England Journal of Medicine* 302:743–45.

Anonymous. 1994. Editorial: changes in semen and the testis. *British Medical Journal* 309:1316.

———. 1996a. Hormonal sabotage. *Natural History* 105(March):42–49.

———. 1996b. Endocrine disruptor research planned by White House, agencies, industry. *Environmental Science & Technology* 30(June):242A–43A.

———. 1996c. Another chemicals balls-up? *Chemistry & Industry* no. 9, May 6, 315.

Apfel, R. J., and S. M. Fisher. 1984. *To Do No Harm.* New Haven: Yale University Press.

Arizona Republic. 1995. Lesbian leanings, estrogen tied. Daughters of moms who took DES also show bisexuality. February 1, A2.

Arnold, S. F., D. M. Klotz, B. M. Collins, et al. 1996. Synergistic activation of estrogen receptor with combinations of environmental chemicals. *Science* 272:1489–92.

Ashby, J., E. Houthoff, S. J. Kennedy, et al. 1997a. The challenge posed by endocrine disrupting chemicals. *Environmental Health Perspectives* 105:164–9.

Ashby, J., P. A. Lefevre, J. Odum, et al. 1997b. Synergy between synthetic estrogens? *Nature* 385:494.

Ashford, Nicholas. 1994. Personal communication. August.

———. 1997. The policy framework for endocrine-disrupting chemicals. *Human Environment* 4(summer):6–7.

Atlanta Journal and Constitution. 1995. Falling sperm count sounds warning. February 6, A8.

Baden, J., and D. S. Noonan. 1996. Al Gore's newest horror story (review of *Our Stolen Future*). *American Enterprise* 7:83–84.

Bailey, R. 1996. Hormones and humbug: a new exposé is one part pseudo-science, two parts hype, three parts hysteria. *Washington Post,* March 31, C3.

Bantle, J., W. W. Bowerman, C. Carey, et al. 1995. Consensus statement from the Work Session on "Environmentally Induced Alterations in Development: A Focus on Wildlife." *Environmental Health Perspectives* 103(suppl. 4):3–5.

Barish, G. 1998. Interview. January 31.

Barnard, J. 1994. Canaries in the coal mine? Wild birds' deformities could have ominous implications for humans. *Salt Lake Tribune,* August 18, C4.

Barnett, J. B., T. Colborn, M. Fournier, et al. 1996. Consensus statement from the Work Session on "Chemically Induced Alterations in the Developing Immune System: The Wildlife/Human Connection." *Environmental Health Perspectives* 104(suppl. 4):807–8.

Beatty, Jack. 1996. Review of *Our Stolen Future. Audubon* 98:112–13.

Begley, S., and D. Glick. 1996. The estrogen complex. *Newsweek,* March 21, 76–77.

Beil, L. 1993. Toxic chemicals' role in breast cancer studied. *Dallas Morning News,* November 1, D8.

Benson, W. H., H. A. Bern, B. Bue, et al. 1997. Consensus statement from the Work Session on "Chemically Induced Alterations in the Functional Development and Reproduction of Fishes." In *Chemically Induced Alterations in the Development and Reproduction of Fishes,* ed. R. Rolland, M. Gilbertson, and R. B. Peterson, 3–8. Pensacola, Fla.: SETAC Press.

Bern, H. A., P. Blair, S. Brasseur, et al. 1992. Consensus statement from the Work Session on "Chemically Induced Alterations in Sexual Development: The Wildlife/Human Connection." In *Chemically Induced Alterations in Sexual and Functional Development: The Wildlife/Human Connection,* ed. T. Colborn and C. Clement, 1–8. Advances in Modern Environmental Toxicology, Vol. 21. Princeton, N.J.: Princeton Scientific.

Bibbo, M., W. B. Gill, F. Azizi, et al. 1977. Follow-up study of male and female offspring of DES-exposed mothers. *Obstetrics and Gynecology* 49:1–8.

Bonner, J. 1996. Review of *Our Stolen Future. New Scientist* 150:46–47.

Bradlow, H. L. 1997. Interview. June 16.

Breyer, S. 1993. *Breaking the Vicious Circle.* Cambridge, Mass.: Harvard University Press.

Brock, J., T. Colborn, R. Cooper, et al. 1999. Consensus statement from the Work Session on "Health Effects of Contemporary-Use Pesticides: The Wildlife/Human Connection." *Toxicology and Industrial Health* 15:1–5.

Brody, J. E. 1996. Study finds lasting damage from prenatal PCB exposure. *New York Times,* September 12, A14.

Broton, J. A., M. F. Olea-Serrano, M. Villalobos, et al. 1995. Xenoestrogens released from lacquer coatings in food cans. *Environmental Health Perspectives* 103:608–12.

Brouwer, A., C. Colborn, M. C. Fossi, et al. 1998. Consensus statement from the Work Session on "Environmental Endocrine-Disrupting Chemicals: Neural, Endocrine, and Behavioral Effects." *Toxicology and Industrial Health* 14:1–8.

Burger, A. 1996. Sex offenders. *E. Magazine,* April, 44–47.

Burlington, H., and V. F. Lindeman. 1950. Effect of DDT on testes and secondary sex characteristics of white leghorn cockerels. *Proceedings of the Society for Experimental Biology and Medicine* 74:48–51.

Cadbury, D. 1997. *The Feminization of Nature.* London: Hamish Hamilton.

Cameron, J., and J. Aboucher. 1991. The precautionary principle: a fundamental principle of law and policy for the protection of the global environment. *Boston College International Comparative Law Review* 14:1–27.

Carey, J. 1996. Review of *Our Stolen Future. Business Week,* April 18, 18.

Carlsen, E., A. Giwervman, N. Keiding, and N. E. Skakkebaek. 1992. Evidence for decreasing quality of semen during the past 50 years. *British Medical Journal* 305:609–13.

Carpenter, B. 1996. Investigating the next silent spring? *U.S. News & World Report,* March 11, 50.

Carson, R. 1962. *Silent Spring.* New York: Fawcett Crest.

Chemical Industry Institute of Toxicology. 1997. Synergism of weakly estrogenic chemicals not confirmed (news release). May 30.

Chen, Y. C. J., Y. L. Guo, C. C. Hsu, et al. 1992. Cognitive development of Yu-cheng (oil-disease) children prenatally exposed to heat-degraded PCBs. *Journal of the American Medical Association* 268:3213–18.

Chen, Y. C. J., M. L. M. Yu, W. J. Rogan, et al. 1994. A 6-year follow-up of behavior and activity disorders in the Taiwan Yu-Cheng children. *American Journal of Public Health* 84:415–21.

Churchman, C. W. 1947. *Theory of Experimental Inference.* New York: Macmillan.

Clorfene-Casten, L. 1993. The environmental link to breast cancer. *Ms.,* May/June, 52–56.

Cohrssen, J. J., and V. T. Covello. 1989. *Risk Analysis: A Guide to Principles and Methods for Analyzing Health and Environmental Risks.* Washington, D.C.: Council on Environmental Quality.

Colborn, T. 1981. Aquatic insects as measures of trace element presence: cadmium and molybdenum. Master's thesis. Western State College of Colorado, Department of Biology.

———. 1985. The use of the stonefly, *Pteronarcys californica* Newport, as a measure of bioavailable cadmium in a high-altitude river system, Gunnison County, Colorado. Ph.D. dissertation. University of Wisconsin, Department of Zoology.

———. 1991. Nontraditional evaluation of risk from fish contaminants. In *Proceedings of a Symposium on Issues in Seafood Safety,* ed. F. E. Ahmed. Washington, D.C.: National Academy of Sciences, Institute of Medicine, Food and Nutrition Board, 95–122.

———.1995. Interview. December 29.

Colborn, T., and C. Clement, eds. 1992. *Chemically Induced Alterations in Sexual and Functional Development: The Wildlife/Human Connection.* Advances in Modern Environmental Toxicology, Vol. 21. Princeton, N.J.: Princeton Scientific.

Colborn, T. E., A. Davidson, S. N. Green, et al. 1990. *Great Lakes, Great Legacy?* Washington, D.C.: Conservation Foundation.

Colborn,T., F. S. vom Saal, and A. M. Soto. 1993. Developmental effects of endocrine disrupting chemicals in wildlife and humans. *Environmental Health Perspectives* 101:378–83.

Colborn, T., D. Dumanoski, and J. P. Myers. 1996. *Our Stolen Future.* New York: Dutton.

Commoner, B. 1971. *Science and Survival.* New York: Viking.

Cone, M. 1994a. Sexual confusion in the wild (The gender war: are chemicals blurring sexual identities?). *Los Angeles Times,* October 2 (first in a series).

———. 1994b. Pollution's effect on sexual development fires debate (The gender war: are chemicals blurring sexual identities?). *Los Angeles Times,* October 3 (second in a series).

———. 1994c. Battle looms on chemicals that disrupt hormones (The gender war: are chemicals blurring sexual identities?). *Los Angeles Times,* October 4 (third and last in a series).

Cortese, A. D. 1996. Endocrine disruption (review of *Our Stolen Future*). *Environmental Science & Technology* 30:213A.

Cottrell, D. 1971. The price of beef. *Environment* 13:44–51.

Cranor, C. F. 1993. *Regulating Toxic Substances.* New York: Oxford University Press.

Crumley, B., L. Mondi, U. Plon, and L. H. Towle. 1996. What's wrong with our sperm? *Time,* March 18, 78.

Cushman, J. H. 1994. E.P.A. Settles Suit and Agrees to Move Against 36 Pesticides. *New York Times,* October 13, A24.

Daly, H. B. 1991. Reward reductions found more aversive by rats fed environmentally contaminated salmon. *Neurotoxicology and Teratology* 13:449–53.

———. 1993. Laboratory rat experiments show consumption of Lake Ontario salmon causes behavioral changes: support for wildlife and human research results. *Journal of Great Lakes Research* 19:784–88.

Daly, H. B., D. R. Hertler, and D. M. Sargent. 1989. Ingestion of environmentally contaminated Lake Ontario salmon by laboratory rats increases avoidance of unpredictable aversive nonreward and mild electric shock. *Behavioral Neuroscience* 103:1356–65.

Danish Environmental Protection Agency. 1995. *Male Reproductive Health and Environmental Chemicals with Estrogenic Effects.* Miljø projekt nr. 290. Copenhagen: Ministry of Environment and Energy.

Dansereau, C. 1996. Pollution: placing children in harm's way. *Seattle Times,* March 20, B5.

Danzo, B. J. 1997. Environmental xenobiotics may disrupt normal endocrine function by interfering with physiological ligands to steroid receptors and binding proteins. *Environmental Health Perspectives* 105:294–301.

Davis, D. L., and H. L. Bradlow. 1995. Can environmental estrogens cause breast cancer? *Scientific American* 273:166–72.

Davis, D. L., H. L. Bradlow, M. Wolff, et al. 1993. Medical hypothesis: xenoestrogens as preventable causes of breast cancer. *Environmental Health Perspectives* 101:372–77.

Davis, D. L., D. Axelrod, M. P. Osborne, et al. 1997. Environmental influences on breast cancer. *Science and Medicine* 4:56–63.

Demarest, H. E. 1996. A Great Lakes time bomb the cleanup crew missed. *San Francisco Chronicle,* April 14, Book Review, 10.

Dodds, E. C., and W. Lawson. 1936. Synthetic oestrogenic agents without the phenanthrene nucleus. *Nature* 137:996.

Dodds, E. C., and W. Lawson. 1938. Molecular structure in relation to oestrogenic activity: compounds without a phenanthrene nucleus. *Proceedings of the Royal Society of London, Series B* 125:222–32.

Dodds, E. C., L. Goldberg, W. Lawson, et al. 1938. Estrogenic activity of certain synthetic compounds. *Nature* 141:247–48.

Dopyera, C. 1994. Chemicals tinker with sexuality. [Raleigh, N.C.] *News and Observer,* September 25, A1, A14.

Dorfman, R. 1982. The lessons of pesticide regulation. In *Reform of Environmental Regulation,* ed. W. A. Magat. Cambridge, Mass.: Ballinger.

Dowie, M. 1998. What's wrong with the *New York Times* science reporting? *Nation,* July 6, 13–14, 16–19.

Duax, W. L., D. C. Swenson, P. D. Strong, et al. 1984. Molecular structures of metabolites and analogues of diethylstilbestrol and their relationship to receptor binding and biological activity. *Molecular Pharmacology* 26:520–25.

Dumanoski, Dianne. 1996/97. Charting the territory of collaboration. *Antioch Journal* 5:40–44.

———.1998. Personal communication, June.

Efron, E. 1984. *The Apocalyptics.* New York: Simon and Schuster.

Ehrlich, P., and A. Ehrlich. 1996. *Betrayal of Science and Reason.* Washington, D.C.: Island Press.

Endocrine Disrupter Screening and Testing Advisory Committee, Environmental Protection Agency. 1998. Final report. Washington, D.C.: Environmental Protection Agency.

Environmental Defense Fund and R. H. Boyle. 1980. *Malignant Neglect.* New York: Vintage.

Environmental Media Services. 1996. Exposure to environmental chemicals: PR hype or public health concern (audiotape recording). June 12.

Environmental Protection Agency, Office of Pollution Prevention and Toxic Substances. 1996. *EPA Activities on Endocrine Disrupters: Background Paper.* Prepared for the meeting "Endocrine Disruption by Chemicals: Next Steps in Chemical Screening and Testing," May 15–16, 1996, Washington, D.C.

Environmental Protection Agency, Technical Panel, Office of Research and Development, Office of Prevention, Pesticides and Toxic Substances. 1997. *Special Report on Environmental Endocrine Disrupters: An Effects Assessment and Analysis.* EPA/630/ R-961012.Washington, D.C.: U.S. Environmental Protection Agency.

Erikson, J. 1995. Breast cancer activists seek voice in research decisions. *Science* 269:1508–9.

Fagin, D., and M. Lavelle. 1996. *Toxic Deception.* Secaucus, N.J.: Carol.

Fairley, P., M. Roberts, and J. Stringer. 1996. Endocrine disrupters: sensationalism or science? *Chemical Week,* May 8, 29.

Falk, F. Y., A. Ricci, M. Wolff, et al. 1992. Pesticides and polychlorinated biphenyl residues in human breast lipids and their relation to breast cancer. *Archives of Environmental Health* 47:143–46.

Fantle, W. 1994. The incredible shrinking man. *The Progressive* 58(October):12–13.

Field, B., M. Selub, and C. I. Hughes. 1990. Reproductive effects of environmental agents. *Seminars in Reproductive Endocrinology* 8:44–54.

Fisch, H., and E. T. Goluboff. 1996. Geographic variations in sperm counts: a potential cause of bias in studies of semen quality. *Fertility and Sterility* 65:1044–46.

Foster, P. 1997. Assessing the effects of chemicals on male reproduction: lessons learned from Di-n-butylphthalate. *CIIT Activities* 17:1–8.

Foulkes, R. G. 1996. Review of *Our Stolen Future. Fluoride* 29:227–29.

Fox, G. A. 1992. Epidemiological and pathobiological evidence of contaminant-induced alterations in sexual development in free-living wildlife. In *Chemically Induced Alterations in Sexual and Functional Development: The Wildlife/Human Connection,* ed. T. Colborn and C. Clement, 147–58. Advances in Modern Environmental Toxicology, Vol. 21. Princeton, N.J.: Princeton Scientific.

———.1996. Interview. February 12.

Fox, M. 1996. Sex chemicals. *Harper's Bazaar,* February, 92, 94.

Frith, M. 1996. Fertility fear after baby milk chemical find. *Press Association Newsfile,* May 25.

Fumento, M. 1996. Sperm care may just be soybeans. *Sacramento Bee,* March 31, F3.

Fung, V. A., J. C. Barrett, and J. Huff. 1995. The carcinogenesis bioassay in perspective: application in identifying human cancer hazards. *Environmental Health Perspectives* 103:680–83.

Funtowicz, S. O., and J. R. Ravetz. 1992. Three types of risk assessment and the emergence of post-normal science. In *Social Theories of Risk,* ed. S. Krimsky and D. Golding, 251–73. New York: Praeger.

Gellert, R. J. 1978. Kepone, mirex, dieldrin, and aldrin: estrogenic activity and the induction of persistent vaginal estrus and anovulation in rats following neonatal treatment. *Environmental Research* 16:131–38.

Gibbons, A. 1991. Reproductive toxicity: regs slow to change. *Science* 254:25.

Giddens, A. 1996. Why sounding the alarm on chemical contamination is not necessarily alarmist (review of *Our Stolen Future*). *London Review of Books* 18:19–20.

Gilbertson, M. 1996. Interview. March 15.

Gillesby, B. E., and T. R. Zacharewski. 1998. Exoestrogens: mechanisms of action and strategies for identification and assessment. *Environmental Toxicology and Chemistry* 17:3–14.

Ginsberg, J. 1996. Tackling environmental endocrine disrupters. *Lancet* 347:1501.

Gladen, B. C., W. J. Rogan, P. Hardy, et al. 1988. Development after exposure to polychlorinated biphenyls and dichlorodiphenyl dichloroethane transplacentally and through breast milk. *Journal of Pediatrics* 113:991–95.

Glas, J. P. 1989. Protecting the ozone layer: a perspective from industry. In *Technology and Environment,* ed. J. H. Ausubel and H. E. Sladovich, 137–55. Washington, D.C.: National Academy Press.

Golder, D. J. 1996. Letter to the editor: exploiting breast cancer won't cure it. *New York Times,* October 7, A16.

Goldman, L. R. 1997. Letter to EDSTAC members, April 24. http://www.epa.gov/opptintr/opptendo/consensu.txt.

Goodman, W. 1994. Something is attacking male fetus sex organs. *New York Times,* September 2, D17.

Gorbach, S. L., D. Zimmerman, and M. Woods. 1984. *The Doctor's Anti–Breast Cancer Diet.* New York: Simon and Schuster.

Gore, A. 1992. *Earth in the Balance.* Boston: Houghton Mifflin.

Graham, F. 1970. *Since Silent Spring.* Boston: Houghton Mifflin.

Greenpeace International. 1995. Body of evidence: the effects of chlorine on human health. May. Public information document.

Guillette, L. J. 1994. Testimony before U.S. Congress, House, Committee on Energy and Commerce, Subcommittee on Health and the Environment. *Health Effects of Estrogenic Pesticides.* Published proceedings. Washington, D.C.: U.S. Government Printing Office.

Hammond, B., B. S. Katzenellenbogen, N. Krauthammer, and J. McConnell. 1979. Estrogenic activity of the insecticide chlordecone (kepone) and interaction with uterine estrogen receptors. *Proceedings of the National Academy of Sciences USA* 76:6641–45.

Hauser, P., A. J. Zametkin, P. Martinez, et al. 1993. Attention deficit–hyperactivity disorder in people with generalized resistance to thyroid hormone. *New England Journal of Medicine* 328:997–1000.

Haybron, R. 1995. Fertility in males is declining. [Cleveland] *Plain Dealer,* February 7, E7.

Healy, M. 1993. Pesticides may be linked to breast cancer, scientists warn. *Los Angeles Times,* October 22, A20.

Hentschel, E., G. Brandstatter, B. Dragosics, et al. 1993. Effect of ranitidine and amoxicillin plus metronidazole on the eradication of *Helicobacter pylori* and the recurrence of duodenal ulcer. *New England Journal of Medicine* 328:308–12.

Herbst, A. L., and H. A. Bern, eds. 1981. *Developmental Effects of Diethylstilbestrol (DES) in Pregnancy.* New York: Thieme-Stratton.

Herbst, A. L., and R. E. Scully. 1970. Adenocarcinoma of the vagina in adolescence: a report of 7 cases including 6 clear cell carcinoma so-called mesomephromas. *Cancer* 25:745–57.

Herbst, A. L., H. Ulfelder, and D. C. Peskanzer. 1971. Adenocarcinoma of the vagina: association of maternal stilbestrol therapy with tumor appearances in young women. *New England Journal of Medicine* 284:878–81.

Hertsgaard, M. 1996. A world awash in chemicals (review of *Our Stolen Future*). *New York Times Book Review,* April 7, 25.

Hileman, B. 1996. Environmental hormone disruptors focus of major research initiatives. *Chemical and Engineering News,* May 13, 28–32.

Hiltbrand, D. 1994. Picks & pans: tube. *People,* September 5, 13.

Hirshfield, A. N., M. F. Hirshfield, and J. A. Flaws. 1996. Problems beyond pesticides (review of *Our Stolen Future*). *Science* 272:1444–45.

Hook, G. E. R., and G. W. Lucier. 1997. Editorial: synergy, antagonism, and scientific process. *Environmental Health Perspectives* 105:784.

Houghton, J. T., G. J. Jenkins, and J. J. Ephraums, eds. 1990. *Climate Change: The IPCC Scientific Assessment.* Cambridge: Cambridge University Press.

Howard, V. 1997. Synergistic effects of chemical mixtures—can we rely on traditional toxicology? *Ecologist* 27:192–94.

Høyer, A. P., P. Grandjean, T. Jørgensen, et al. 1998. Organochlorine exposure and risk of breast cancer. *Lancet* 352:1816–20.

Huber, P. 1991. *Galileo's Revenge: Junk Science in the Courtroom.* New York: Basic Books.

Hunt, V. R., K. M. Smith, D. Worth, et al., eds. 1982. *Environmental Factors in Human Growth and Development.* Cold Spring Harbor, N.Y.: Cold Spring Harbor Laboratory.

Hunter, D. J., S. E. Hankinson, F. Laden, et al. 1997. Plasma organochlorine levels and the risk of breast cancer. *New England Journal of Medicine* 337:1253–58.

Hynes, H. P. 1989. *The Recurring Silent Spring.* Elmsford, N.Y.: Pergamon Press.

Illinois Environmental Protection Agency. 1997. *Endocrine Disruptors Strategy.* February.

International Agency for Research on Cancer. 1991. *Monographs on the Evaluation of Carcinogenic Risks to Humans: Occupational Exposure in Insecticide Application, and Some Pesticides.* Lyons: IARC.

International Organization for Economic Cooperation and Development. 1997. European Workshop on the Impact of Endocrine Disrupters on Human Health and Wildlife. Report of proceedings. DGX11, April 16. EUR 17459.

Jacobson, J. L., and S. W. Jacobson. 1996. Intellectual impairment in children exposed to polychlorinated biphenyls in utero. *New England Journal of Medicine* 335:783–89.

Jacobson, J. L., S. W. Jacobson, and H. E. B. Humphrey. 1990. Effects of in utero exposure to polychlorinated biphenyls and related contaminants on cognitive functioning in young children. *Journal of Pediatrics* 116:36–45.

Jacobson, S. W., G. G. Fein, J. L. Jacobson, et al. 1985. The effect of intrauterine PCB exposure on visual recognition memory. *Child Development* 56:853–60.

Johnson, H. D. 1996. A disturbing sequel to *Silent Spring. San Francisco Examiner,* May 5, C15.

Johnson, J. 1996. Endocrine disruption. *Environmental Science & Technology* 30:168A–70A.

Johnstone, J. W. 1997. Editorial: combating junk science. *Chemical and Engineering News,* April 28, 5.

Kaiser, J. 1996. Endocrine disrupters: Scientists angle for answers. *Science* 274:1837–38.

Kamrin, M. A. 1996. The mismeasure of risk (review of *Our Stolen Future*). *Scientific American* 275:178–79.

Katz, D. 1997. The press's ignominious role. *Wall Street Journal.* August 20, A14.

Kavlock, R. J., G. P. Daston, C. DeRosa, et al. 1996. Research needs for the risk assessment of health and environmental effects of endocrine disruptors: a report of the U.S. EPA-sponsored workshop. *Environmental Health Perspectives* 104(suppl. 4):715–40.

Kay, J. 1994. Cancer linked to use of DDT. But some dispute Israeli findings. *San Francisco Examiner,* March 2, A4.

Key, T., and G. Reeves. 1994. Organochlorines in the environment and breast cancer. *British Medical Journal* 308:1520–21.

Kolata, G. 1996a. Chemicals that mimic hormones spark alarm and debate. *New York Times,* March 19, C1, C10.

———. 1996b. Sperm counts: some experts see a fall, others poor data. *New York Times,* March 19, C10.

———. 1996c. Are U.S. men less fertile? Latest research says no. *New York Times,* April 29, A4.

Krieger, N., M. S. Wolff, R. A. Hiatt, et al. 1994. Breast cancer and serum organochlorines: a prospective study among white, black, and Asian women. *Journal of the National Cancer Institute* 86:589–99.

Krimsky, S. 1982. *Genetic Alchemy: The Social History of the Recombinant DNA Controversy.* Cambridge, Mass.: MIT Press.

———. 1992. The role of theory in risk studies. In *Social Theories of Risk,* ed. S. Krimsky and D. Golding. New York: Praeger.

Krimsky, S., and D. Golding, eds. 1992. *Social Theories of Risk.* New York: Praeger.

Krimsky, S., and A. Plough. 1988. *Environmental Hazards: Communicating Risks as a Social Process.* Dover, Mass.: Auburn House.

Krimsky, S., and L. S. Rothenberg. 1998. Financial interest and its disclosure in scientific publications. *Journal of the American Medical Association* 280:1–2.

Krimsky, S., J. Ennis, and R. Weissman. 1991. Academic-corporate ties in biotechnology: a quantitative study. *Science, Technology, and Human Values* 16:275–87.

Krimsky, S., L. S. Rothenberg, P. Stott, and G. Kyle. 1996. Financial interests of authors in scientific journals: a pilot study of 14 publications. *Science and Engineering Ethics* 2:395–410.

Kuhn, T. 1962. *The Structure of Scientific Revolutions.* Chicago: University of Chicago Press.

Lahita, R. G., H. L. Bradlow, H. G. Kunkel, and J. Fishman. 1981. Increased 16-alpha-hydroxylation of estradiol in systemic lupus erythematosus. *Journal of Clinical Endocrinology and Metabolism* 53:174–78.

Lamb, J. C. 1997. Can today's risk assessment paradigms deal with endocrine active chemicals? *Risk Policy Report* 4:30, 32–33.

Langley, J. W. 1874. Synthetic chemistry. *Popular Science Monthly* 5:39–46.

Lave, L. B., and A. C. Upton. 1987. Regulating toxic chemicals in the environment. In *Toxic Chemicals, Health, and the Environment,* ed. L. B. Lave and A. C. Upton, 280–93. Baltimore: Johns Hopkins University Press.

Lear, L. 1997. *Rachel Carson: Witness for Nature.* New York: Holt.

Leary, W. E. 1998. Research ties radon to as many as 21,800 deaths each year. *New York Times,* February 20, A13.

Lee, G. 1996. Poisoned planet. *Washington Post,* April 14, X9.

Lemonick, M. D. 1996. Not so fertile ground. *Time,* September 19, 68–70.

Lewis, S. 1990. Federal statutes. In *Fighting Toxics,* ed. G. Cohen and J. O'Connor, 165–208. Washington, D.C.: Island Press.

Lucier, G. W., and G. E. R. Hook. 1996. Anniversaries and issues. *Environmental Health Perspectives* 104:350.

Luoma, J. R. 1992a. New effect of pollutants: hormone mayhem. *New York Times,* March 24, C1.

———. 1992b. Cancer not only contaminant concern: hormonal systems profoundly affected. *San Diego Union Tribune,* April 1, C3.

Macilwain, C. 1998. U.S. panel split on endocrine disruptors. *Nature* 397:828.

McKenna, M. A. J. 1994. Could men become extinct? Pesticides and plastics may threaten male sex hormones. *Boston Herald,* July 14, 1, 39.

McKinney, J. D. 1997. Editorial: interactive hormonal activity of chemical mixtures. *Environmental Health Perspectives* 105:896–97.

McLachlan, J. A., ed. 1980. *Estrogens in the Environment.* New York: Elsevier North-Holland.

———, ed. 1985. *Estrogens in the Environment II: Influences on Development.* New York: Elsevier North-Holland.

———.1995. Interview. May 22.

———.1997. Letter: synergistic effects of environmental estrogens: report withdrawn. *Science* 277:462–63.

McLachlan, J. A., and K. S. Korach, eds. 1995. *Estrogens in the Environment III: Global Health Implications.* Washington, D.C.: National Institutes of Health, National Institute of Environmental Health Sciences.

McLachlan, J. A., and R. R. Newbold. 1987. Estrogens and development. *Environmental Health Perspectives* 75:25–27.

Malkin, M. 1996. A technophobe's whimper about the end of the world. *Seattle Times,* March 12, B4.

Mathews, J. 1996. Overlooking the "POPs" problem. *Washington Post,* March 11, A19.

Meadows, D. 1996. A chemical whirlwind on the horizon (review of *Our Stolen Future*). *Los Angeles Times,* January 31, B9.

Meilahn, E. N., B. De Stavola, D. S. Allen, et al. 1998. Do urinary oestrogen metabolites predict breast cancer? Guernsey III cohort follow-up. *British Journal of Cancer* 78:1250–55.

Moghissi, A. A. 1996. Is the future stolen? (review of *Our Stolen Future*). *Environment International* 22:275–77.

Monmaney, T. 1993. Marshall's hunch: annals of medicine. *New Yorker,* September 20, 64–72.

Moomaw, W. 1996. Hormone mimics in the environment (review of *Our Stolen Future*). *Chemical & Engineering News,* April 1, 34–35.

Myers, J. P. 1998. Interview. February 16.

———.1999. Interview. February 24.

Myers, N. 1979. *The Sinking Ark.* New York: Pergamon.

National Academy of Sciences, Commission on Life Sciences. 1994. *Statement of Contract: Hormone-Related Toxicants in the Environment.* September 13. Washington, D.C.: National Academy of Sciences.

National Research Council. 1983. *Risk Assessment in the Federal Government: Managing the Process.* Washington, D.C.: National Academy Press.

———.1999. *Hormonally Active Agents in the Environment.* Washington, D.C.: National Academy Press.

National Wildlife Federation. 1994a. *Hormone Copy Cats.* Unpublished report of the Great Lakes Natural Resources Center, National Wildlife Federation. April 4. Washington, D.C: National Wildlife Federation.

———.1994b. *Fertility on the Brink: The Legacy of the Chemical Age.* Washington, D.C: National Wildlife Federation.

Natural Resources Defense Council. 1989. *Intolerable Risk: Pesticides in Our Children's Food.* Unpublished report. February 27. New York: Natural Resources Defense Council.

Needleman, H. L., and C. A. Gatsonis. 1990. Low-level lead exposure and the IQ of children: a meta-analysis of modern studies. *Journal of the American Medical Association* 263:673–78.

Newbold, R. R., and J. A. McLachlan. 1996. Transplacental hormonal carcinogenesis: diethyl stilbestrol as an example. In *Cellular and Molecular Mechanisms of Hormonal Carcinogenesis,* ed. J. Juff, J. Boyd, and J. C. Barrett. New York: Wiley-Liss.

Newbold, R. R., B. C. Bullock, and J. A. McLachlan. 1990. Uterine adenocarcinoma in mice following developmental treatment with estrogens: a model for hormonal carcinogenesis. *Cancer Research* 50:7677–81.

New York Times. 1988. Editorial: worry chic. January 7, A26.

———. 1996. New study questions radon danger in houses. July 17, A15.

Osborn, F. 1948. *Our Plundered Planet.* Boston: Little, Brown.

Osborne, M. P., H. L. Bradlow, G. Y. Wang, et al. 1993. Increase in the extent of estradiol 16α-hydroxylation in human breast tissue: a potential biomarker of breast cancer risk. *Journal of the National Cancer Institute* 85:1917–20.

Palmlund, I. 1996. Exposure to xenoestrogens before birth: the diethylstilbestrol experience. *Journal of Psychometrics, Obstetrics, and Gynecology* 17:71–84.

Patlak, M. 1996. A testing deadline for endocrine disrupters. *Environmental Science and Technology/News* 30:542.

Pauling, L. 1958. *No More War!* New York: Dodd, Mead.

Paulsen, C. A., N. C. Berman, and C. Wang. 1996. Data from men in greater Seattle area reveals no downward trend in semen quality: further evidence that deterioration of semen quality is not geographically uniform. *Fertility and Sterility* 65:1015–20.

Perrings, C. 1991. Reserved rationality and the precautionary principle: technological change, time and uncertainty in environmental decision making. In *Ecological Economics,* ed. R. Costanza, 153–66. New York: Columbia University Press.

Peterson, K. 1993. Decreasing sperm counts blamed on the environment. *USA Today,* May 28, A1.

Piller, C., and K. Yamamoto. 1988. *Gene Wars.* New York: William Morrow.

Pinchbeck, D. 1996. Downward motility. *Esquire* 125(January):78–84.

Planck, M. 1949. *Scientific Autobiography and Other Papers.* New York: Philosophical Library.

Popper, K. 1959. *The Logic of Scientific Discovery.* New York: Harper and Row.

Portney, K. E. 1992. *Controversial Issues in Environmental Policy.* Newbury Park, Calif.: Sage.

Raeburn, P. 1996. From silent spring to barren spring (review of *Our Stolen Future*). *Business Week,* March 18, 42.

Raloff, J. 1993. Plastics may shed chemical estrogens. *Science News* 144:12.

———. 1994a. The gender benders: are environmental "hormones" emasculating wildlife? *Science News* 145:24–27.

———. 1994b. Estrogen's malevolence: that feminine touch. Are men suffering from prenatal or childhood exposures to "hormonal" toxicants? *Science News* 145:56–59.

Ramamoorthy, K., F. Wang, I. C. Chen, et al. 1997. Estrogenic activity of a dieldrin/toxaphene mixture in the mouse uterus, MCF-7 human breast cancer cells, and yeast-based estrogen receptor assays: no apparent synergism. *Endocrinology* 138:1520–27.

Reijnders, P. J. H., and S. M. J. M. Brasseur. 1992. Xenobiotic induced hormonal and associated developmental disorders in marine organisms and related effects on humans: an overview. In *Chemically Induced Alterations in Sexual and Functional Development: The Wildlife/Human Connection,* ed. T. Colborn and C. Clement, 159–74. Advances in Modern Environmental Toxicology, Vol. 21. Princeton, N.J.: Princeton Scientific.

Richmond Times-Dispatch. 1996. Editorial: polluting the debate. March 31, F6.

Rivero-Rodriguez, L., V. H. Borja-Aburto, C. Santos-Burgoa, et al. 1997. Exposure assessment for workers applying DDT to control malaria in Veracruz, Mexico. *Environmental Health Perspectives* 105:98–101.

Roan, S. 1989. *Ozone Crisis.* New York: John Wiley and Sons.

Roberts, M. 1996. U.K. health scare over phthalates in infant formula. *Chemical Week,* June 5, 22.

Rogan, W. J., B. C. Gladen, K. L. Hung, et al. 1988. Congenital poisoning by polychlorinated biphenyls and their contaminants in Taiwan. *Science* 241:334–36.

Rolland, R., M. Gilbertson, and R. B. Peterson, eds. 1997. *Chemically Induced Alterations in the Development and Reproduction of Fishes.* Pensacola, Fla.: SETAC Press.

Rosen, J. D. 1990. Much ado about Alar. *Issues in Science and Technology* 7:85–90.

Roush, W. 1995. Conflict marks crime conference. *Science* 269:1808–9.

Rudolph, J. 1999. Coming to the defense of the human guinea pig. *Boston Globe,* January 17, C3.

Safe, S. H. 1994. Dietary and environmental estrogens and antiestrogens and their possible role in human disease. *Environmental Science and Pollution Research International* 1:29–33.

———. 1995a. Environmental and dietary estrogens and human health: is there a problem? *Environmental Health Perspectives* 103:346–51.

———.1995b. Environmental estrogens—response. *Environmental Health Perspectives* 103:784–85.

———. 1997a. Editorial: another enviro-scare debunked. *Wall Street Journal,* August 20, A14.

———. 1997b. Editorial: xenoestrogens and breast cancer. *New England Journal of Medicine* 337:1303–4.
———.1997c. Interview. April 16.
Safe, S. H., and A. McDougal. 1997. Environmental factors and breast cancer. *Endocrine-Related Cancer* 4:1–11.
Safe, S. H., and K. Ramamoorthy. 1998. Disruptive behavior. *Forum* 13(Fall):19–23.
Schettler, Ted. 1998. Interview. April 13.
Schneider, J., D. Kinne, A. Fracchia, et al. 1982. Abnormal oxidative metabolism of estradiol in women with breast cancer. *Proceedings of the National Academy of Sciences USA* 79:3047–51.
Scialli, A. R. 1996. The developmental toxicity of the H-1 histamine antagonists. *Reproductive Toxicology* 10:247–55.
Sever, L., T. E. Arbuckle, and A. Sweeney. 1997. Reproductive and developmental effects of occupational pesticide exposure: the epidemiological evidence. *Occupational Medicine* 12:305–25.
Sharpe, R. M., and N. E. Skakkebaek. 1993. Are oestrogens involved in falling sperm counts and disorders of the male reproductive tract? *Lancet* 431:1392–95.
Sheehan, D. M., and F. S. vom Saal. 1997. Low dose effects of endocrine disruptors: a challenge for risk assessment. *Inside EPA's Risk Policy Report* 4:31, 35–39.
Shrader-Frechette, K. 1991. *Risk and Rationality.* Berkeley: University of California Press.
———.1994. *Ethics of Scientific Research.* London: Rowman and Littlefield.
Silent Spring Institute. 1996. Report to the Public Advisory Committee for the Cape Cod Breast Cancer and Environmental Study, Section II. Newton, Mass.
Silverstein, K. 1996. APCO: Astroturf makers. *Multinational Monitor,* March, 28.
Skakkebaek, N. E. 1996. Interview. November 5.
Skakkebaek, N. E., A. Negro-Vilar, and F. Michal, eds. 1993. Proceedings of the International Workshop on the Impact of the Environment on Reproductive Health, September 30–October 4, 1991. *Environmental Health Perspectives* 101(suppl. 2):1–167.
Snow, T. 1996. Book raises alarm about human survival. *Hattiesburg American,* March 21.
Soto, A. M., T. M. Lin, J. H. Silva, et al. 1992. An "in-culture" bioassay to assess the estrogenicity of xenobiotics (E-SCREEN). In *Chemically Induced Alterations in Sexual and Functional Development: The Wildlife/Human Connection,* ed. T. Colborn and C. Clement, 295–309. Advances in Modern Environmental Toxicology, Vol. 21. Princeton, N.J.: Princeton Scientific.
Soto, A. M., C. Sonnenschein, K. L. Chung, et al. 1995. The E-SCREEN assay as a tool to identify estrogens: an update on estrogenic environmental pollutants. *Environmental Health Perspectives* 103(suppl. 7):113–22.
Springston, R. 1996. Authors issue new pesticide warnings. *Richmond Times Dispatch,* April 14, A1, A10.
Steingraber, S. 1997a. *Living Downstream.* Reading, Mass.: Addison-Wesley.
———. 1997b. Mechanism, proof, and unmet needs: the perspective of a cancer activist. *Environmental Health Policy* 105(suppl. 3):685–87.
Stevens, W. K. 1994. Pesticides may leave legacy of hormonal chaos. *New York Times,* August 23, C1, C6.
Stillman, R. J. 1982. In utero exposure to diethylstilbestrol: adverse effects on the reproductive tract and reproductive performance in male and female offspring. *American Journal of Obstetrics and Gynecology* 142:905–21.
Stoll, B. A. 1969. *Hormonal Management in Breast Cancer.* Philadelphia: Lippincott.
Stone, R. 1994. Environmental estrogens stir debate. *Science* 265:308–10.
Sullivan, L. W. 1996. Chemical villains (review of *Our Stolen Future*). *Los Angeles Times,* April 1, B5.

Suominen, J., and M. Vierula. 1993. Semen quality of Finnish men. *British Medical Journal* 306:1579.

Swan, S. H., E. P. Elkin, and L. Fenster. 1997. Have sperm densities declined? A reanalysis of global trend data. *Environmental Health Perspectives* 105:128–32.

Swan, S. H., K. Waller, B. Hopkins, and G. DeLorenze. 1998. Trihalomethanes in drinking water and spontaneous abortion. *Journal of Epidemiology* 9:134–40.

Taylor, R. 1996. Ecologist fears a future stolen by man-made chemicals. *Seattle Post-Intelligencer,* April 2, C1.

Telang, N. T., F. Arcuri, D. M. Granata, et al. 1998. Alteration of estradiol metabolism in *myc* oncogene–transfected mouse mammary epithelial cells. *British Journal of Cancer* 77:1549–54.

Thrupp, L. A. 1991. Sterilization of workers from pesticide exposure: the causes and consequences of DBCP-induced damage in Costa Rica and beyond. *International Journal of Health Services* 21:731–57.

Toppari, J., J. C. Larsen, P. Christiansen, et al. 1996. Male reproductive health and environmental xenoestrogens. *Environmental Health Perspectives* 104(suppl. 4):741–76.

Tye, L. 1998. Journal fuels conflict-of-interest debate. *Boston Globe,* January 6, B1, B8.

Unger, M., H. Kiaer, M. Blichert-Toft, et al. 1984. Organochlorine compounds in human breast fat from deceased with and without breast cancer and in a biopsy material from newly diagnosed patients undergoing breast surgery. *Environmental Research* 34:24–28.

U.S. Agency for Toxic Substances and Disease Registry, Department of Health and Human Services, Public Health Service. 1998. *Public Health Implications of Persistent Toxic Substances in the Great Lakes and St. Lawrence Basins.* Atlanta: Department of Health and Human Services.

U.S. Congress, House, Committee on Agriculture. 1996. 104th, 2nd session. *Food Quality Protection Act of 1996 (HR1627).* Report of the committee. July 8. Washington, D.C.: U.S. Government Printing Office.

U.S. Congress, House, Committee on Commerce. 1996. 104th, 2nd session. *Food Quality Protection Act of 1996 (HR1627).* Report of the committee. July 23. Washington, D.C.: U.S. Government Printing Office.

U.S. Congress, House, Committee on Commerce, Subcommittee on Health and the Environment. 1996. 104th, 1st session. Hearings: *Food Quality Protection Act of 1995.* June 7, 29, 1995. Washington, D.C.: U.S. Government Printing Office.

U.S. Congress, House, Committee on Energy and Commerce, Subcommittee on Health and the Environment. 1994. 103rd, 1st session. *Health Effects of Estrogenic Pesticides.* Published proceedings. Washington, D.C.: U.S. Government Printing Office.

U.S. Congress, House, Committee on Energy and Commerce, Subcommittee on Health and the Environment, and Senate Committee on Labor and Human Resources. 1994. 103rd, 1st session. Joint hearings: *Safety of Pesticides in Food.* September 21, 1993. Washington, D.C.: U.S. Government Printing Office.

U.S. Congress, Senate. 1995. 104th, 1st session. *Congressional Record* 141(189):S17749 (November 29).

U.S. Congress, Senate, Committee on Governmental Affairs. 1992. 102nd, 1st session. Hearings: *Government Regulation of Reproductive Hazards,* October 2, 1991. Washington, D.C.: U.S. Government Printing Office.

U.S. General Accounting Office. 1990. *Toxic Substances: EPA's Chemical Testing Program Has Made Little Progress.* April 13. Washington, D.C.: U.S. Government Printing Office.

———. 1992. *Reproductive and Developmental Toxicants: Regulatory Actions Provide Uncertain Protection.* October. Washington, D.C.: U.S. Government Printing Office.

Van, S. 1999. Researchers at Cedars-Sinai Medical Center to present first documentation of man-made chemical contaminants in the amniotic fluid of unborn babies (Cedars-Sinai Medical Center news release). June 14.

Vom Saal, F. S. 1996. Interview. February 13.

———.1997. Interview. October 31.

Vom Saal, F., and F. Bronson. 1980. Sexual characteristics of adult female mice are correlated with their blood testosterone levels during prenatal development. *Science* 208:597–99.

Vom Saal, F. S., and W. V. Welshons. 1997. Letter to Lynn Harris, Society of the Plastics Industry. June 12.

Vom Saal, F. S., W. Grant, C. McMullen, and K. Laves. 1983. High fetal estrogen titres correlate with enhanced adult sexual preferences and decreased aggression in male mice. *Science* 220:1306–9.

Vom Saal, F. S., P. S. Cooke, D. L. Buchanan, et al. 1998. A physiologically based approach to the study of bisphenol-A and other estrogenic chemicals on the size of reproductive organs, daily sperm production, and behavior. *Toxicology and Industrial Health* 14:239–60.

Waldholz, M. 1996. Scientists debate the future threat of common chemicals. *Wall Street Journal,* March 7, B9, B14.

Wapner, K. 1995. Chemical sleuth: Theo Colborn studies waterways and wildlife. *Amicus Journal* 17:18–21.

Wargo, J. 1996. *Our Children's Toxic Legacy.* New Haven, Conn.: Yale University Press.

Washington Times. 1996. Editorial: sperm limits. March 13, A18.

Weiser, P., R. Muller, U. Braun, and M. Roth. 1997. Endosomal targeting and the cytoplasmic tail of membrane immunoglobin: retraction. *Science* 277:20–21.

Weiss, R. 1994. Estrogens in the environment: are some pollutants a threat to fertility? *Washington Post,* January 25, Z10.

———.1997. Tulane researchers retract findings on pollutants' risk. *Washington Post,* August 17, A15.

Weiss, R., and G. Lee. 1996. Pollution's effect on human hormones: when fear exceeds evidence. *Washington Post,* March 31, A14.

Whelan, E. M., W. M. London, and L. T. Flynn. 1996. *ACSH Commentary on Our Stolen Future.* Washington, D.C.: American Council on Science and Health.

Whorton, M. D., and T. H. Milby. 1980. Recovery of testicular function among DBCP workers. *Journal of Occupational Medicine* 22:177–79.

Whorton, M. D., T. H. Milby, R. M. Krauss, and H. A. Stubbs. 1979. Testicular function in DBCP-exposed pesticide workers. *Journal of Occupational Medicine* 21:161–66.

Wiles, R. 1994. Testimony: U.S. Congress, 103rd, 1st session, Subcommittee on Health and the Environment, Committee on Energy and Commerce, Hearings: *Health Effects of Estrogenic Pesticides.* October 21.

Wilson, P. W. 1940. *The Biochemistry of Symbiotic Nitrogen Fixation.* Madison: University of Wisconsin Press.

Windsor, Jr., A. S. 1996. Another eco alarm. *The New American,* April 29, 31–32.

Wolff, M. S., and P. J. Landrigan. 1994. Letter. *Science* 266:525–26.

Wolff, M. S., P. G. Toniolo, E. W. Lee, et al. 1993. Blood levels of organochlorine residues and risk of breast cancer. *Journal of the National Cancer Institute* 85:648–52.

Worldwatch Institute. 1994. *State of the World 1994.* New York: W. W. Norton.

Wright, L. 1996. Silent sperm. *The New Yorker,* January 15, 42.

Yang., R. S. H. 1996. Some current approaches for studying combination toxicology in chemical mixtures. *Food & Chemical Toxicology* 34:1037–44.

Zeeman, M. 1996. Our fate is connected with the animals (review of *Our Stolen Future*). *BioScience* 46:542–44.

Zilinskas, R. A., and B. K. Zimmerman. 1986. *The Gene Splicing Wars.* New York: Macmillan.

Zimmerman, M. 1996. Toxic bodies. *Philadelphia Inquirer,* March 24, K1, K6.

Zinberg, D. S. 1997. Editorial: a cautionary tale. *Science* 273:411.

INDEX

References to figures are followed by an *f* and those to tables by a *t*.